S0-CDQ-479

Communication Complexity

ACM Doctoral Dissertation Awards

1982

Area-Efficient VLSI Computation

Charles Eric Leiserson

1983

Generating Language-Based Environments

Thomas W. Reps

1984

Reduced Instruction Set Computer Architectures for VLSI

Manolis G. H. Katevenis

1985

Bulldog: A Compiler for VLIW Architectures

John R. Ellis

1986

Computational Limitations for Small Depth Circuits

Torkel Hastad

Full Abstraction and Semantic Equivalence

Ketan Mulmuley

1987

The Complexity of Robot Motion Planning

John Canny

1988

Communication Complexity: A New Approach to Circuit Depth

Mauricio Karchmer

Communication Complexity: A New approach to Circuit Depth

Mauricio Karchmer

The MIT Press
Cambridge, Massachusetts
London, England

This book was printed and bound in the United States of America

Library of Congress Cataloging-in-Publication Data

Karchmer, Mauricio.
Communication complexity : a new approach to circuit depth / Mauricio Karchmer.
p. cm. -- (ACM doctoral dissertation awards :)
Bibliography: p.
Includes index.
ISBN 0-262-11143-8
1. Boolean algebra. 2. Logic circuits. 3. Computational complexity. 4. Automatic theorem proving. I. Title. II. Series: ACM doctoral dissertation award.
QA10.3.K37 1989
511.3'24--dc20 89-34499
CIP

To my Parents
Samuel and Susana

Contents

Series Foreword

This volume is the doctoral thesis of Mauricio Karchmer, winner of the 1988 Distinguished Doctoral Dissertation Award, sponsored by the Association of Computing Machinery and The MIT Press.

The award, first presented in 1982, identifies the best doctoral thesis in computer science and engineering submitted in the past year. The award has become very widely known and very competitive. This year there were more than 50 nominees, 9 of whom were from school outside the United States. Dr. Karchmer's thesis, submitted by Hebrew University, is the first foreign winner.

Writing under the direction of Professor Avi Wigderson, Dr. Karchmer studies the circuit complexity of Boolean functions, a topic of long-standing interest in computer science. Claude Shannon was the first to observe that although most Boolean functions must be complex, we do not know any specific examples. This problem of finding lower bounds on the complexity of circuits for Boolean functions has motivated much research over the past 40 years, but it has proved to be very stubborn.

In this thesis Karchmer develops a new strategy for studying circuit complexity, which he calls the communication complexity approach. This approach casts the computational device as a mechanism for *separating* words from the nonwords of language, and it explicitly captures the notion of the *flow of information*. Apploying this reformulation to monotone circuits, he shows a tight $Q(\log^2 n)$ monotone depth bound for st-connectivity, the problem of determining the existence of a path between distinguished vertices s and t in an undirected graph. This implies a superpolynomial ($n^{w(\log n)}$) lower bound for monotone circuits computing st-connectivity.

Lawrence Snyder
Chair, ACM Distingished Doctoral Dissertation
Award Selection Committee

Preface

In this thesis we propose a new approach to study the depth of boolean circuits: The *Communication Complexity approach.* The approach is based on an equivalence between the circuit depth of a given function, and the communication complexity of a related problem. The bottom-line of the new approach is that it looks at a computation device as a *separating* device; that is, a device that separates the words of a language from the non-words. This allows us to view computation in a *Top-Down* fashion and makes explicit the idea that *flow of information* is a crucial term for understanding computation.

We demonstrate that the communication complexity approach is both useful and intuitive. We do so by

- Giving new simpler proofs to old results which help us understand the results in the correct setting.
- Proving a super-logarithmic monotone depth lower bound for the function st-connectivity.

We present, in our new setting, results of Berkowitz and Dunne relating monotone and non-monotone computation. Also, we present upper bounds to some functions by giving *protocols* for the communication problems associated with them, and we introduce the notion of universal relations which, in a sense, correspond to universal circuits.

What best exemplifies the first item above is, perhaps, a new proof of a theorem of Khrapchenko. The original proof of the theorem gave no clue whatsoever to the fact that its truth stems from a simple information theoretic fact: *One needs* $\log d$ *bits to distinguish among* d *possibilities.*

We present a tight $\Theta(\log^2 n)$ depth lower bound for monotone circuits computing the function *st-connectivity*, a function which has $O(n^3 \log n)$ size monotone circuits. This is our main technical contribution: A monotone depth lower bound which is *super-logarithmic* in the size of the best circuit for the function considered. That is, our techniques apply to depth rather than to size. Thus, our results complement those by Andreev and Razborov who obtained exponential size lower bounds for monotone circuits computing some functions in NP. As a consequence, we get both super-polynomial ($n^{\Omega(\log n)}$) size lower bounds for monotone formulas computing *st*-connectivity, and a separation of the monotone analogues of NC^1 and AC^1.

Acknowledgments

First, I would like to thank my advisor Avi Wigderson. He honored me, from the first day, by treating me more as a colleague than as a student: This thesis is all joint work with him. More than that, and above all, he has always been a very good friend.

I am grateful to Allan Borodin for sharing with me his love for, and thoughts about, circuit complexity. I have, also, learned so much from Michael Ben-Or, Nati Linial and Eli Shamir.

I have no complaints whatsoever about my fellow students: Hagit Attiya, Judit Bar-Ilan, Amotz Bar-Noy, Yosi Ben-Asher, Aviad Cohen, Seffi Naor, Ilan Newman, Tal Rabin, Nir Shavit and Mike Werman. We have always shared much more than just scientific problems.

I am grateful to Valerie King for helping me get rid of some of my grammatical errors, misplaced commas and bad spelling.

Finally, I thank (if that is the word) my wife Evelyn. If I have worked in peace during these years, it is mostly because of her being.

Communication Complexity:
A New approach to Circuit Depth

Chapter 1

Introduction

The complexity of Boolean functions has been studied for almost 40 years. The field has developed into a theory in this, perhaps short, period mainly because of the success in defining both a set of complexity measures (those for circuit complexity and for Turing machine complexity) and a surprising hierarchy of very robust complexity classes. Moreover, characteristics of most of the defined classes have been understood by showing complete problems for them. Relations between some of the complexity classes have been discovered, and new models have been developed. The main frustration of the theory has been, however, the inability of showing a separation of any two classes (excluding those obtained by diagonalization methods*). To state it simply, the main problem remains unsolved: Though it is known that most functions are complex [Sh49], we do not have an example of a simple function (say in NP) that requires super-linear circuit size, or super-logarithmic circuit depth.

The reason for our inability to obtain non-trivial lower bounds is, perhaps, that although the circuit model is elegantly simple, our understanding of the way it computes is, at most, vague. There seems to be a need to develop more intuitive ways of looking at computation. A new approach may give some clues as to where to look for the heart of complexity and, at the same time, shed some light on how to prove lower bounds.

In this thesis we would like to propose a new approach to circuit depth: The *Communication Complexity approach* †. The approach is based on an equivalence between the circuit depth of a given function, and the communication complexity of a related problem. The bottom-line of the new approach is that it looks at

*Diagonalization methods are not strong enough to separate such classes as P and NP (see [BGS75]).

†Yannakakis independently discovered this equivalence which is implicit in [KPPY].

a computation device as a *separating* device; that is, a device that separates the words of a language from the non-words. The characterization of circuit depth in terms of communication complexity is reminiscent of, but somehow more explicit and intuitive than, the well-known relationship between circuits and alternating machines [Ru80]. Among other things, the new approach allows us to view computation in a *top-down* fashion. Also, the approach makes explicit the idea that *flow of information* is a crucial term for understanding computation.

We will demonstrate that the communication complexity approach is very intuitive, and that it captures, in a strong way, the essence of circuit depth. We will do so by:

- Giving new, simpler proofs to old results which become clearer in this new setting.
- Proving a super-logarithmic monotone depth lower bound for the function st-connectivity.

In 1985, work of Andreev [An85] and Razborov [Ra85a], later improved by Alon and Boppana [AB], lead to exponential monotone size lower bounds for such functions in NP as $CLIQUE$. These results separate the monotone analogues of P and NP. Though these results can be used to obtain exponential (in $\log n$) monotone depth lower bounds as well, the depth lower bound is always logarithmic in the size bound. That is, the techniques apply to size rather than to depth. Our contribution is to present monotone depth lower bounds which are super-logarithmic in the size of the best circuit for the function considered. In this way, our results complement those by Andreev and Razborov. We present a tight $\Theta(\log^2 n)$ depth bound for st-connectivity [‡], a function which has $O(n^3 \log n)$ size monotone circuits.

[‡] We present an improved and simplified version of an early result giving a $\Omega(\log^2 n/\log\log n)$ bound. This was possible after J. Hastad and, independently, R. Boppana formulated and proved lemma 5.1.1.

As a consequence, we get both a super-polynomial ($n^{\Omega(\log n)}$) size lower bound for monotone formulas computing *st*-connectivity, and a separation of the monotone analogues of NC^1 and AC^1.

This thesis is organized as follows:

In § 2, we give an overview of the relevant definitions and previous work of both circuit complexity and communication complexity. In this chapter we treat these fields as two unrelated ones. We present a slightly different treatment to communication complexity from that in the literature. The main difference is that we consider mainly search problems, as opposed to decision problems.

In § 3, we develop our main thesis by defining and proving the equivalence between circuit depth (or formula size) and a related search problem in communication complexity. In this chapter we also vary the search problem in order to capture the essence of monotone circuit depth. We finish the chapter by giving some general consequences of the new approach.

In § 4, we demonstrate that the communication complexity approach is very intuitive by *i)* Presenting new, more intuitive proofs for some old results; and *ii)* Defining some new concepts which come about naturally in the communication approach. In § 4.1, we present new proofs of some results concerning slice functions, and the relation between monotone and non-monotone computation. In § 4.2, we show that the new approach may help us, not only to understand better some known upper bounds, but also to improve upon the known ones. In this section we present a couple of such examples. In § 4.3, we introduce the concept of a universal relation (closely related to that of a universal circuit). We give both deterministic as well as randomized protocols for these universal relations. We also show that, while the universal relation has efficient randomized protocols, its monotone version does not. Finally, in § 4.4, we present a new proof of a depth analogue of a theorem of Khrapchenko. We believe that this example best exemplifies the power of the new setting.

In § 5, we demonstrate the usefulness of the new approach by presenting two monotone depth lower bounds. In § 5.1 we present the depth lower bound for *st*-connectivity. This is our main technical contribution. We would like to emphasize that most of the ideas behind the proof, and even the flow of the argument, were suggested by the new approach. In § 5.2, we present a recent result of Razborov [Ra88] which uses communication complexity to give a monotone lower bound for *MINIMUM COVER*.

In our last chapter, § 6, we comment upon some points regarding the approach in general, and our proofs in particular. We also propose some open problems which, we feel, will lead the way towards proving a general depth lower bound.

Preliminary results from this work have been published in [KW88]. The material contained in § 5.2 did not appear in the thesis of the author but is included in order to make this work more complete.

Chapter 2

Definitions and Previous Work

Before going into our main thesis, let us review both Circuit Complexity and Communication Complexity as two unrelated topics.

2.1 Circuit Complexity

2.1.1 Definitions

A *Boolean Circuit* is a directed acyclic graph with each node of indegree either 0 or 2 and a single node of outdegree 0. Nodes of indegree 0 are called *inputs* and are labeled by either a variable x_i or its negation $\bar{x}_i$. Nodes of indegree 2 are called *gates* and are labeled by either of the Boolean operations $\{\wedge, \vee\}$. The single node of outdegree 0 is called the *output* of the circuit. A *Boolean Formula* is a Boolean circuit where each node, other than the output, has outdegree 1. Boolean circuits and formulae compute Boolean functions in a natural way.

For a circuit C, its *size*, $s(C)$, is defined as the number of edges it contains. Its *depth*, $d(C)$, is the maximum distance from an input to the output. For a formula F, its *size*, $L(F)$, is the number of input nodes. For a function f, we define $s(f)$ as the minimum size of a circuit computing f. Similarly, we define $d(f)$ and $L(f)$.

A *monotone Boolean function* f is such that $x \leq y^{\dagger}$ implies $f(x) \leq f(y)$. For a monotone function f, a *minterm* (*maxterm*) is a minimal set of variables which if we set to 1 (0), the function f is set to 1 (0). Let $min(f)$, $Max(f)$ be the set of minterms and maxterms of f respectively. The following fact is immediate:

†For $x, y \in \{0,1\}^n$, $x \leq y$ iff. $\forall i\ x_i \leq y_i$.

Fact 2.1.1 *Let f be a monotone function. For every $p \in min(f)$ and every $q \in Max(f)$, $p \cap q \neq \emptyset$.*

A *monotone circuit* is a circuit where no input node is labeled by a negated variable $\bar{x}_i$. We define monotone formulae in an analogous way. It is well-known that monotone circuits and formulae compute precisely monotone functions. For a monotone circuit C, a monotone formula F, and a monotone function f, we define $s_m(C)$, $d_m(C)$, $L_m(F)$, $s_m(f)$, $d_m(f)$, and $L_m(f)$ in the obvious way.

The following facts are well-known both for the general and the monotone complexities.

Fact 2.1.2 $s(f) \leq L(f)$.

Fact 2.1.3 $d(f) = \Theta(\log L(f))$.

Fact 2.1.3 says that, up to constant factors and a different scale, circuit depth and formula size represent similar complexity measures. In this work, we will be concerned mainly with circuit depth, and thus, with formula size.

The definitions of Boolean circuits can be extended to allow unbounded fanin gates. An *Unbounded Fanin Circuit* is similar to a Boolean circuit with the difference that gates have unbounded indegree and are labeled by unbounded fanin $\wedge$'s and $\vee$'s. Size and depth are defined in the same way as for general circuits.

To study asymptotic complexity, we define families of functions and families of circuits computing them. A *family* $\mathcal{F}$ of functions is a sequence $f_1, f_2, \ldots$, where f_n is a function of $g_{\mathcal{F}}(n)$ variables for some $g_{\mathcal{F}} : N \mapsto N$ with $g_{\mathcal{F}}(n) \geq n$. A *family* $\mathcal{C}$ of circuits is a sequence $C_1, C_2, \ldots$, where C_n is a Boolean circuit on $g_{\mathcal{C}}(n)$ variables for some $g_{\mathcal{C}} : N \mapsto N$ with $g_{\mathcal{C}}(n) \geq n$. We say that $\mathcal{C}$ computes $\mathcal{F}$ if, for all n,

$g_{\mathcal{F}}(n) = g_{\mathcal{C}}(n)$ and C_n computes f_n. We naturally extend the notion of a family of circuits to define families of formulae and unbounded fanin circuits.

Having these notions of families we can proceed to define (non-uniform) complexity classes.

Definition 2.1.1 *Let $Depth(g(n))$ be the set of all families of functions $\mathcal{F}$ such that, for all n, $d(f_n) = O(g(n))$.*

Size related classes (e.g. *size*, *formula-size*) can be defined in a similar way.

Definition 2.1.2 *NC^k is the set of all families $\mathcal{F}$ for which there exist a family $\mathcal{C}$ of circuits computing $\mathcal{F}$ such that, for all n, $d(C_n) = O(\log^k(n))$ and $s(C_n) = n^{O(1)}$.*

Definition 2.1.3 *AC^k is the set of all families $\mathcal{F}$ for which there exist a family $\mathcal{C}$ of unbounded fanin circuits computing $\mathcal{F}$ such that, for all n, $d(C_n) = O(\log^k(n))$ and $s(C_n) = n^{O(1)}$.*

Definition 2.1.4 *P is the set of all families $\mathcal{F}$ for which there exist a family $\mathcal{C}$ of circuits computing $\mathcal{F}$ such that, for all n, $s(C_n) = n^{O(1)}$.*

Definition 2.1.5 *PL is the set of all families $\mathcal{F}$ for which there exist a family $\mathcal{C}$ of formulae computing $\mathcal{F}$ such that, for all n, $L(C_n) = n^{O(1)}$.*

Denote $NC = \cup_k NC^k$ and $AC = \cup_k AC^k$. The following facts are well-known:

Fact 2.1.4 $PL = NC^1$.

Fact 2.1.5 *For every k, $AC^k \subseteq NC^{k+1} \subseteq AC^{k+1}$.*

Fact 2.1.6 $NC = AC \subseteq P$.

Monotone versions of the above classes are defined accordingly and differentiated by a subindex m (e.g. NC_m). All three of the above facts hold also for the monotone cases.

2.1.2 Previous Work

We will concentrate on previous work related to either circuit depth or formula size in order to locate our results in the appropriate context.

Though it is known that almost every function has depth $n - \Theta(\log n)$ [Sh49], it is still a major challenge in circuit complexity to construct an explicit function (say in NP) with depth $\omega(\log n)$. The best general lower bound for formula size is contained in the work of Khrapchenko [K71] which will be presented, with a new proof, in section 4.4. The best bound attainable with this method is $L(\oplus_n) \geq n^2$ where $\oplus_n$ is the n-bit parity function. Better formula size lower bounds have been proven recently by Andreev [An86] who constructed a particular family $\mathcal{G}$ of functions with $L(g_n) = \Omega(n^{5/2-\epsilon})$. This still gives a depth lower bound of the form $\Omega(\log n)$.

Because of the lack of understanding of general circuit complexity, people turned to restricted models as research subjects with the goal of gaining intuition for the general models. By far, the most important restrictions considered have been depth bounded circuits and monotone circuits.

In 1981, Ajtai [Aj83] and independently Furst, Saxe and Sipser [FSS84] proved perhaps the first nontrivial lower bound. They showed that unbounded fanin circuits of polynomial size require nonconstant depth to compute $\oplus_n$, and thus $AC^0 \subset NC^1$. In 1986, Yao [Y85] and Hastad [H86] actually showed that exponential size is necessary for constant depth circuits computing $\oplus_n$, or that if we bound the size by a polynomial, these circuits require depth $\Omega(\log n / \log\log n)$. All these results are based on a technique which uses *random restrictions* to show, in a *bottom-up* fashion†, that small shallow circuits compute very simple functions.

In 1985, work by Andreev [An85] and Razborov [Ra85a], improved by Alon and Boppana [AB87], led to exponential ($2^{\Omega(n^{\epsilon})}$) monotone size lower bounds for such functions in NP as $CLIQUE$, and thus $P_m \subset NP_m$. This, using fact 2.1.3, gives monotone depth lower bounds of the form $\Omega(n^{\epsilon})$. Note that, although the depth lower bound is super-logarithmic in n, it is still logarithmic in the size of the circuit. These results were obtained using a technique termed *circuit approximation.* It is worthwhile to note that, though this technique is different from that for unbounded fanin circuits, it still uses a *bottom-up* argument to show that small circuits can be approximated well by simple functions.

In § 5.1, we will give a tight monotone depth lower bound for st-connectivity of the form $\Omega(\log^2 n)$. This shows that $NC_m^1 \subset AC_m^1$. By fact 2.1.3, we get that st-connectivity requires monotone formula size $n^{\Omega(\log n)}$. We note that there exists monotone circuits for st-connectivity of size $O(n^3 \log n)$ so that the depth lower bound is super-logarithmic in the size of the circuit, i.e. the depth is non-trivial. In other words, the result shows that, for monotone computation, restrictions on the outdegree can blow the size up from polynomial to super-polynomial. The proof uses a *top-down* argument which shows that, if the output of the circuit is in a sense complex, there is always a node below it which computes a complex subfunction. A similar top-down argument has been used before in [KPPY84] to prove the existence of a depth-hierarchy within AC_m^0.

It is interesting to note here the different character of the connectivity and majority functions in the Boolean and arithmetic monotone circuits models. Shamir and Snir [ShS80] proved an $\Omega(\log^2 n)$ depth bound for both functions in the arithmetic

---

†We think of a circuit with inputs at the bottom and the output at the top. In this way, restrictions are applied to the inputs.

model[‡]. The difficulty in applying these techniques to Boolean circuits resides in the axioms $x \vee xy = x$ and its dual, which do not hold in rings. Indeed, Valiant [V84] (by probabilistic methods) and [AKS83] (by explicit constructions) showed that these axioms make a difference for the majority function which admits monotone Boolean circuits of depth $O(\log n)$. Our result says that, unlike for majority, for connectivity the situation in the Boolean case is very similar to that in the arithmetic one.

Recently, Razborov [Ra88] obtained monotone lower bounds of the form $\Omega(\log^2 n)$ for the function $MINIMUM\ COVER$. Though the results stated above can be used to obtain a lower bound for this function of the form $\Omega(n^c)$, the new proof uses entirely new methods. In particular, it uses rank arguments to show, in a *global* way, the desired bound. This has to be contrasted to other results that use either *bottom-up* or *top-down* arguments. In § 5.2, we will present a communication complexity version of this result.

2.2 Communication Complexity of Relations

2.2.1 Definitions

We consider the following scenario: Two players, I and II, both with unlimited computing power, communicate through a flawless binary channel. The players follow a deterministic protocol, and we require that they communicate using prefix-free codes so that at each point of time, history uniquely determines the players' turn to send a message.

Consider three finite sets X, Y and Z, and a ternary relation $R \subseteq X \times Y \times Z$.

[‡]A monotone arithmetic circuit is a circuit with gates labeled by $\{+, \times\}$ and inputs labeled by either variables, or constant from some field F. These circuits compute, in a natural way, a polynomial in $F[x_1, ..., x_n]$.

Let $S(R) \subseteq X \times Y$, the *support* of R, be the set of pairs $(x, y) \in X \times Y$ such that there exists a $z \in Z$ with $(x, y, z) \in R$. We say that a relation $R \subseteq X \times Y \times Z$ is *rectangular* if $S(R) = X \times Y$.

Given a relation R, consider the following *game* between players I and II: For $(x, y) \in X \times Y$, give x to player I and y to player II. Their goal is to agree on *any* $z \in Z$ with the proviso that $(x, y) \in S(R)$ implies that $(x, y, z) \in R$. Let D be a protocol for the above game for R. For $(x, y) \in X \times Y$, we write $D(x, y)$ for the number of bits communicated by the players when they follow D on (x, y). Also, we denote by $\alpha(x, y)$ the *communication pattern* or *history* of D on (x, y).

Definition 2.2.1 *Let $R \subseteq X \times Y \times Z$ be a relation. The* communication complexity *of R, $C(R)$, is defined as*

$$C(R) = \min \max_{(x,y) \in S(R)} D(x, y)$$

where the minimum is taken over all protocols D for R.

Without loss of generality, we can assume that for every history α of D, there exists at least one $(x, y) \in S(R)$ such that $\alpha(x, y) = \alpha$. This means that definition 2.2.1 is equivalent to the one where we maximize over all $(x, y) \in X \times Y$. Also, every $R \subseteq X \times Y \times Z$ can be extended to a relation

$$\bar{R} = R \cup \{(x, y, z) : (x, y) \notin S(R) \text{ and } z \in Z\}$$

in such a way that: *i)* $S(\bar{R}) = X \times Y$; *ii)* if $(x, y) \in S(R)$ then for every $z \in Z$, $(x, y, z) \in R$ iff $(x, y, z) \in \bar{R}$; and *iii)* $C(R) = C(\bar{R})$. We call $\bar{R}$ the *extension (of a relation)* of R. To see *iii)*, note that every protocol D for R serves as a protocol for $\bar{R}$ which is always correct on pairs $(x, y) \in S(R)$ and, by definition, also on pairs not in $S(R)$. We keep this as a fact for later use:

Fact 2.2.1 *For every $R \subseteq X \times Y \times Z$, $C(R) = C(\bar{R})$.*

It is possible to look at the above game from the point of view of a third party. It is easy to see that a third party can infer the answer to the game by just listening to the conversation between I and II, even if he cannot see either x or y. This is because every history α is associated with a cartesian product $X' \times Y'$ where $X' \subseteq X$, $Y' \subseteq Y$ and such that for every $(x, y) \in X' \times Y'$, $\alpha(x, y) = \alpha$. Sometimes it is worthwhile to take this third party's view.

For complexity theoretic purposes, we will be interested in asymptotic complexity. To this end we have the following definition.

Definition 2.2.2 *A* family $\mathcal{R}$ *of relations is a sequence of relations* $R_n \subseteq X_n \times Y_n \times Z_n$ *for* $n = 1, 2, \ldots$.

We can now define complexity classes in a natural way:

Definition 2.2.3 *Let* $Comm(g(n))$ *be the set of all families of relations* $\mathcal{R}$ *such that, for all* n, $C(R_n) = O(g(n))$.

In what follows, we will describe a family $\mathcal{R}$ by its generic element R_n and call it R if no confusion arises.

Randomized and Distributional Complexities. We next define randomized and distributional communication complexities.

Definition 2.2.4 *Let* $R \subseteq X \times Y \times Z$. *A* randomized *protocol for* R *is a probability distribution over all protocols for* $X \times Y \times Z$.

Note that the above definition is equivalent to the one based on a protocol between two probabilistic players where random bits can be communicated for free, i.e., one where the players share a common random source. Also note that in definition 2.2.4 we consider, not only protocols for R, but all protocols for $X \times Y \times Z$. This allows us to have one notion of randomized protocol for two different complexity measures. Indeed, given $R \subseteq X \times Y \times Z$, one can think of two natural complexity measures for randomized protocols. The first requires the protocol to give always correct answers and measures the average communication complexity. The second allows mistakes but bounds the complexity of the worst input.

For a randomized protocol $\tilde{D}$ for R and $(x, y) \in X \times Y$, let $\tilde{D}(x, y)$ be the average number of bits (over $\tilde{D}$) communicated by the players on input (x, y).

Definition 2.2.5 *Let $R \subseteq X \times Y \times Z$ be a relation. The* Las-Vegas communication complexity *of R, $LV\text{-}C(R)$, is defined as*

$$LV\text{-}C(R) = \min \max_{(x,y) \in S(R)} \tilde{D}(x, y)$$

where the minimum is taken over all randomized protocols for R such that only protocols for R have positive probability (i.e., protocols that do not make mistakes).

Let $R \subseteq X \times Y \times Z$. A *Monte-Carlo* protocol $\tilde{D}$ for R is a randomized protocol such that, for every $(x, y) \in S(R)$, the probability that $\tilde{D}$ gives an answer z such that $(x, y, z) \in R$ is at least $1 - 1/n$. For such a protocol, its complexity is the maximum, over all $(x, y) \in S(R)$ and all protocols with positive probability, of $D(x, y)$.

Definition 2.2.6 *Let $R \subseteq X \times Y \times Z$. The* Monte-Carlo Communication Complexity *of R, $MC\text{-}C(R)$, is the minimum complexity of a Monte-Carlo protocol for R.*

Let λ be a probability distribution over $S(R)$. For a deterministic protocol D for

R, let D_λ be the average communication complexity of D with respect to λ, i.e.

$$D_\lambda = \sum_{(x,y) \in S(R)} \lambda(x,y) \cdot D(x,y)$$

Definition 2.2.7 *Let $R \subseteq X \times Y \times Z$ be a relation and let λ be a probability distribution over $S(R)$. The* distributional communication complexity *of R with respect to λ, $C_\lambda(R)$, is defined as*

$$C_\lambda(R) = \min_D D_\lambda$$

where the minimum is taken over all deterministic protocols for R.

Yao [Y81] observed that the *min-max* theorem (Von-Neumann) can be applied to these situations to obtain, in our case, the following relation between Las-Vegas complexity and distributional complexity:

Theorem 2.2.1 *Let $R \subseteq X \times Y \times Z$ be a relation. Then*

$$LV\text{-}C(R) = \max_\lambda C_\lambda(R)$$

where the maximum is taken over all probability distributions λ on $S(R)$.

Restrictions and Reductions. In studying the complexity of a relation R, it may be helpful to restrict it to a subset of $S(R)$. We have the following definition:

Definition 2.2.8 *Let $R \subseteq X \times Y \times Z$ be a relation and $I \subseteq S(R)$. The* restriction *of R into I, $R|_I$, is the relation $\{(x,y,z) \in R : (x,y) \in I\}$.*

We have the following easy, but very helpful lemma:

Lemma 2.2.1 *For a relation $R \subseteq X \times Y \times Z$ and $I \subseteq S(R)$ we have $C(R|_I) \leq C(R)$.*

Proof: Any protocol for R can be used as a protocol for $R|_I$. ∎

We can also define a reducibility notion between relations:

Definition 2.2.9 *Let $R \subseteq X \times Y \times Z$ and $R' \subseteq X' \times Y' \times Z'$. We say that R is* reducible *to R', $R \leq R'$, if there exist functions $\phi_I : X \mapsto X'$, $\phi_{II} : Y \mapsto Y'$ and $\psi : Z' \mapsto Z$ such that for every $(x, y) \in S(R)$:*

1. $(\phi_I(x), \phi_{II}(y)) \in S(R')$ and

2. $(\phi_I(x), \phi_{II}(y), z') \in R' \Rightarrow (x, y, \psi(z')) \in R$.

The motivation for the above definition is contained in the following lemma:

Lemma 2.2.2 *Let $R \subseteq X \times Y \times Z$ and $R' \subseteq X' \times Y' \times Z'$. If $R \leq R'$ then $C(R) \leq C(R')$.*

Proof: We construct a protocol D for R out of a protocol D' for R'. On (x, y), D simulates D' on $(\phi_I(x), \phi_{II}(y))$. If D' answers z' then D answers $\psi(z')$. ∎

Definition 2.2.10 *Let $R \subseteq X \times Y \times Z$ and $R' \subseteq X' \times Y' \times Z'$. We say that R and R' are* equivalent*, $R \equiv R'$, if both $R \leq R'$ and $R' \leq R$.*

It is clear that if $R \equiv R'$ then $C(R) = C(R')$. We can give a slight generalization of definition 2.2.9:

Definition 2.2.11 *Let $R \subseteq X \times Y \times Z$ and $R' \subseteq X' \times Y' \times Z'$. We say that R is* α-reducible *to R', $R \leq_\alpha R'$, if there exist functions $\phi_I : X \mapsto X'$ and $\phi_{II} : Y \mapsto Y'$ such that:*

1. $\forall (x, y) \in S(R)$: $(\phi_I(x), \phi_{II}(y)) \in S(R')$ and

2. $\forall z' \in Z' : C(R|_{I(z')}) \leq \alpha$ where
$I(z') = \{(x,y) \in S(R) : (\phi_I(x), \phi_{II}(y), z') \in R'\}$.

Lemma 2.2.3 *Let $R \subseteq X \times Y \times Z$ and $R' \subseteq X' \times Y' \times Z'$. If $R \leq_\alpha R'$ then $C(R) \leq C(R') + \alpha$.*

Proof: We construct a protocol D for R out of a protocol D' for R'. On (x, y), D simulates D' on $(\phi_I(x), \phi_{II}(y))$. If D' answers z' then D follows a protocol for $R|_{I(z')}$. ∎

Note that $\leq$ coincides with $\leq_0$. To see this, note that if for every $z' \in Z'$, $C(R|_{I(z')}) = 0$, then there exists a $z \in Z$ such that for every $(x, y) \in I(z')$, $(x, y, z) \in R$. We can, then, define $\psi(z')$ to be this z. Also note that if $R_1 \leq_\alpha R_2$ and $R_2 \leq_\beta R_3$ then $R_1 \leq_{\alpha+\beta} R_3$.

2.2.2 Previous Work

This scenario has been studied before in the literature, starting with [Y79], for the case where for every $x \in X$ and $y \in Y$, there is a unique z such that $(x, y, z) \in R$. That is, when R defines a function $r : X \times Y \mapsto Z$. It may be illustrative to take a quick look at this case.

Consider a function $F : X \times Y \mapsto Z$[†] and let $X' \subseteq X$ and $Y' \subseteq Y$. We call the Cartesian product $X' \times Y'$ a *monochromatic rectangle* if F is constant on $X' \times Y'$. A *decomposition* Γ of F is a collection of pairwise disjoint monochromatic rectangles whose union is $X \times Y$. For such a decomposition Γ, its size, $|\Gamma|$, is the number of rectangles it contains. Let $\Psi(F)$ be the minimum size of a decomposition of F. The following theorem of Yao [Y79] is fundamental in proving lower bounds for $C(F)$.

[†] F defines naturally a relation on $X \times Y \times Z$. We abuse notation and call F both the function and the relation.

Theorem 2.2.2 (Yao) $C(F) \geq \log \Psi(F)$.

Example 2.2.1 *Let* $ID_n : \{0,1\}^n \times \{0,1\}^n \mapsto \{0,1\}$ *where for* $x, y \in \{0,1\}^n$, $ID_n(x,y) = 1$ *iff* $x = y$.

Corollary 2.2.1 $C(ID_n) \geq n + 1$.

Proof: It is easy to show that $\Psi(ID_n) = 2^{n+1}$. ∎

In general, $\Psi(F)$ may be hard to compute. The following result of Mehlhorn and Schmidt [MS82] gives a method for bounding $\Psi(F)$ from below. Consider any field K such that $Z \subseteq K$. For any function $F : X \times Y \mapsto Z$, we define the matrix M_F over K with rows and columns indexed by X and Y respectively and whose (x,y)-entry is equal to $F(x,y)$.

Theorem 2.2.3 (Mehlhorn, Schmidt) $\Psi(F) \geq rk(M_F)$.

Example 2.2.2 *Let* $I_{k,n} : [n]^k \times [n]^k \mapsto \{0,1\}$ *where for* $x, y \in [n]^k$, $I_{k,n}(x,y) = 1$ *iff* $x \cap y = \emptyset$‡.

Corollary 2.2.2 $\Psi(I_{k,n}) \geq \binom{n}{k}$.

Proof: It is known that $rk(M_{I_{k,n}}) = \binom{n}{k}$ over the reals [K72]. ∎

For general rectangular relations $R \subseteq X \times Y \times Z$, one can give a natural extension of the above method. The extension, however, won't be so helpful in proving lower bounds for $C(R)$. The difficulty resides in the fact that, for $(x,y) \in X \times Y$, the players may have some freedom in choosing an appropriate $z \in Z$ such that $(x,y,z) \in R$.

‡ $[n] = \{1, ..., n\}$; $[n]^k = \{S \subseteq [n] : |S| = k\}$.

Let $R \subseteq X \times Y \times Z$. A function $F : X \times Y \mapsto Z$ is *consistent* with R, if for every $(x, y) \in X \times Y$, $(x, y, F(x, y)) \in R$, that is, if $F(x, y)$ is a good answer for the communication game for R. Let $\mathcal{F}(R)$ be the collection of all functions consistent with R.

Theorem 2.2.4 *Let $R \subseteq X \times Y \times Z$ be a rectangular relation. Then,*

$$C(R) \geq \log \min_{F \in \mathcal{F}(R)} \Psi(F)$$

In order to use theorem 2.2.4 to give lower bounds for $C(R)$, one has to prove lower bounds for $\Psi(F)$ for a possibly large family of functions F, and not just for one as in corollary 2.2.1. The following extension of the theorem of Mehlhorn and Schmidt reduces the problem to proving lower bounds for the rank of a large family of matrices.

Corollary 2.2.3 *For any field K,*

$$C(R) \geq \log \min_{F \in \mathcal{F}(R)} rk(M_F)$$

Chapter 3

Communication Complexity and Circuit Depth

In this chapter we develop our main thesis: *Communication complexity as a new approach to circuit depth.* The chapter is organized as follows: In § 3.1 we state and prove the equivalence between circuit depth and communication complexity of relations, while in § 3.2 we describe the equivalence for the particular case of monotone circuits. In § 3.3 we present the relation between formula size and communication complexity. In § 3.4 we define synchronized protocols and, finally, in § 3.5 we give some general consequences of the results of this chapter.

3.1 The General Game

Let $B_0, B_1 \subseteq \{0,1\}^n$ be such that $B_0 \cap B_1 = \emptyset$. Consider the rectangular relation $R(B_1, B_0) \subseteq B_0 \times B_1 \times [n]$ where $(x, y, i) \in R(B_1, B_0)$ iff $x_i \neq y_i$, and the game for $R(B_1, B_0)$ as described in § 2.2: Player I gets $x \in B_1$ while player II gets $y \in B_0$; their goal is to agree on a coordinate i such that $x_i \neq y_i$. We write $C(B_1, B_0)$ instead of $C(R(B_1, B_0))$ to denote the minimum number of bits they have to communicate in order for both to agree on such coordinate. For a Boolean function $f : \{0,1\}^n \mapsto \{0,1\}$, we write R_f instead of $R(f^{-1}(1), f^{-1}(0))$. We might also write $R[f]$ instead of R_f if many subindices need to be used. Our main thesis is based on the following theorem which was independently discovered by Yannakakis:

Theorem 3.1.1 *For every function $f : \{0,1\}^n \rightarrow \{0,1\}$ we have*

$$d(f) = C(R_f)$$

Proof: Follows from lemmas 3.1.1 and 3.1.2 below. ▮

Before stating lemmas 3.1.1 and 3.1.2, let us make two little digressions: First, as stated in § 2.2.2, the communication complexity of decision problems has an elegant and fairly well understood theory [Y79],[AUY83], unlike that of search problems which have not been studied before. In view of theorem 3.1.1, we expect that search problems will inherit some of the difficulties of circuit lower bounds. Second, the nonuniformity of the circuit model is mapped, through theorem 3.1.1, to the unlimited computing power of the players. We have not pursued here the possibilities of restricting the power of the players or by requiring uniform players, i.e. resource restricted turing machines. We would like to point out also at the equality sign in theorem 3.1.1; that is, the two quantities in consideration ($d(f)$ and $C(R_f)$) are not only of the same order of magnitude, but they are precisely the same number. Let us now prove theorem 3.1.1.

Lemma 3.1.1 *For all functions $f : \{0,1\}^n \mapsto \{0,1\}$ and all $B_0, B_1 \subseteq \{0,1\}^n$ such that $B_0 \subseteq f^{-1}(0)$ and $B_1 \subseteq f^{-1}(1)$ we have*

$$C(B_1, B_0) \leq d(f)$$

Proof: In words, this lemma shows how to make a protocol for $R(B_1, B_0)$ out of a circuit for f. The players follow down a path of the circuit from the output to one of the inputs whose label defines the answer. They do so in such a way that, at all times, player I has a vector which evaluates to 1 in the current node while player II has a vector which evaluates to 0. Player I simulates the $\vee$-gates along the path, while player II simulates the $\wedge$-gates. In his turn, each player tells the other which input wire (to the present gate) to follow.

Formally, we proceed by induction on $d(f)$.

If $d(f) = 0$ then f is either x_i or $\bar{x}_i$. In either case, we have that for all $x \in B_1$ and $y \in B_0$, $x_i \neq y_i$ so that i is always an answer and $C(B_1, B_0) = 0$.

For the induction step, we suppose that $f = f_1 \wedge f_2$ (the case $f = f_1 \vee f_2$ is treated similarly) and that $d(f) = \max(d(f_1), d(f_2)) + 1$. Let $B_0^j = B_0 \cap f_j^{-1}(0)$ for $j = 1, 2$. By induction, we have that $C(B_1, B_0^j) \leq d(f_j)$ for $j = 1, 2$. Consider the following protocol for B_1 and B_0: II sends a 0 if $y \in B_0^1$, otherwise he sends a 1; the players then follow the best protocol for each of the subcases. We have

$$C(B_1, B_0) \leq 1 + \max_{j=1,2}(C(B_1, B_0^j)) \leq 1 + \max_{j=1,2}(d(f_j)) = d(f)$$

∎

The converse is as follows:

Lemma 3.1.2 *Let $B_0, B_1 \subseteq \{0,1\}^n$ be such that $B_0 \cap B_1 = \emptyset$. Then, there exists a function $f : \{0,1\}^n \mapsto \{0,1\}$ with $B_0 \subseteq f^{-1}(0)$ and $B_1 \subseteq f^{-1}(1)$ such that*

$$d(f) \leq C(B_1, B_0)$$

Proof: In words, the lemma shows how to define a function f, and a circuit computing it, out of a protocol for $R(B_1, B_0)$. One starts by drawing the tree defined by the different histories of the protocol. Then, one labels the inner nodes by $\vee$'s and $\wedge$'s according to who spoke in the corresponding turn. Finally, one labels the leaves with the variable indexed by the answer, and negates it or not according to the induction base described below. The function f is defined as the output of the circuit constructed.

Formally, we proceed by induction on $C(B_1, B_0)$.

If $C(B_1, B_0) = 0$ then there exists an i such that for every $x \in B_1$ and for every $y \in B_0$, $x_i \neq y_i$. It is clear that for every $x', x'' \in B_1$ we have $x'_i = x''_i$ and that the same holds for all $y', y'' \in B_0$. Without loss of generality, $x_i = 1$ so that letting $f = x_i$ we have $B_0 \subseteq f^{-1}(0)$ and $B_1 \subseteq f^{-1}(1)$.

To prove the induction step, we assume that II sends the first bit (the other case is treated similarly). For some partition $B_0 = B_0^1 \cup B_0^2$, II sends a 0 if $y \in B_0^1$, and a 1 otherwise; the players then continue with the best protocol for each of the subcases and

$$C(B_1, B_0) = 1 + \max_{j=1,2}(C(B_1, B_0^j))$$

By induction, there exist f_1, f_2 so that $B_0^j \subseteq f_j^{-1}(0)$, $B_1 \subseteq f_j^{-1}(1)$ and $d(f_j) \leq C(B_1, B_0^j)$ for $j = 1, 2$. Taking now $f = f_1 \wedge f_2$ we have

$$B_1 \subseteq f_1^{-1}(1) \cap f_2^{-1}(1) = f^{-1}(1)$$

$$B_0 = B_0^1 \cup B_0^2 \subseteq f_1^{-1}(0) \cup f_2^{-1}(0) = f^{-1}(0)$$

and

$$d(f) \leq 1 + \max_{j=1,2}(d(f_j)) \leq 1 + \max_{j=1,2}(C(B_1, B_0^j)) = C(B_1, B_0)$$

∎

3.2 The Monotone Game

For monotone circuits we can give a modified version of theorem 3.1.1 which captures, in a nice way, the restrictions of monotone computation.

Let $P, Q \subseteq \mathcal{P}([n])$[†] be such that for every $p \in P$ and for every $q \in Q$, $p \cap q \neq \emptyset$. Consider the rectangular relation $R(P,Q) \subseteq P \times Q \times [n]$ where $(p, q, i) \in R(P, Q)$ iff $i \in p \cap q$. The game for $R(P,Q)$ is, then, as follows: Player I gets a set $p \in P$ while player II gets a set $q \in Q$; their goal is to find an element in $p \cap q$. Once again, we write $C(P,Q)$ instead of $C(R(P,Q))$ to denote the minimum number of bits they have to communicate in order to find such an element. For a monotone function f, we write R_f^m instead of $R(min(f), Max(f))$ (By fact 2.1.1, R_f^m is indeed rectangular). We might also write $R^m[f]$ for R_f^m if too many subindices are used.

[†] $\mathcal{P}([n]) = \{S : S \subseteq [n]\}$.

Also, let $R_f^1 \subseteq f^{-1}(1) \times f^{-1}(0) \times [n]$ where for $x \in f^{-1}(1)$, $y \in f^{-1}(0)$ and $i \in [n]$, $(x, y, i) \in R_f^1$ iff $x_i = 1$ and $y_i = 0$.

Theorem 3.2.1 *For every monotone function f we have*

$$d_m(f) = C(R_f^1) = C(R_f^m)$$

Proof: We first show that $d_m(f) = C(R_f^1)$. Note that, in the base case of lemma 3.1.1, if the circuit is monotone, we always find a coordinate i such that $x_i = 1$ while $y_i = 0$. On the other hand, if the protocol always gives a coordinate i with the above property, lemma 3.1.2 gives a monotone circuit. Since the inductive step is identical , this part is done.

We now finish the theorem by proving that $R_f^1 \equiv R_f^m$:

i) $R_f^m \leq R_f^1$: Let $p \in min(f)$ and $q \in Max(f)$. Let $\phi_I(p)_i = 1$ iff $i \in p$ and let $\phi_{II}(q)_i = 0$ iff $i \in q$. It is clear that $\phi_I(p) \in f^{-1}(1)$ and $\phi_{II}(q) \in f^{-1}(0)$. Furthermore, $(\phi_I(p), \phi_{II}(q), i) \in R_f^1$ iff $i \in p \cap q$ so that we can take $\psi(i) = i$.

ii) $R_f^1 \leq R_f^m$: Let $x \in f^{-1}(1)$ and $y \in f^{-1}(0)$. Let $p_x = \{i : x_i = 1\}$ and $q_y = \{i : y_i = 0\}$. It is easy to see that there exists a $p \subseteq p_x$ and a $q \subseteq q_y$ such that $p \in min(f)$ and $q \in Max(f)$. Furthermore, it is clear that $i \in p \cap q$ iff $x_i = 1$ and $y_i = 0$. We meet the conditions of definition 2.2.9 by taking $\phi_I(x) = p$, $\phi_{II}(y) = q$ and $\psi(i) = i$. ∎

3.3 Communication Complexity and Formula Size

As stated in § 2.1.1, circuit depth and formula size represent similar complexity measures. As it turn out, there is also a very natural interpretation of formula size in terms of communication complexity. In fact, the definition of formula size as the number of leaves of the underlying tree of the formula suggests a similar measure for protocols:

Definition 3.3.1 *Let $R \subseteq X \times Y \times Z$. The* tree complexity *of R, $\Gamma(R)$, is defined as*

$$\begin{aligned} \Gamma(R) &= \min_D \{\# \text{ of histories of } D\} \\ &= \min_D |\{\alpha(x,y) : (x,y) \in S(R)\}| \end{aligned}$$

where the minimum is taken over all protocols for R.

Note that, in definition 3.3.1, we are only counting histories that actually happen. That is, we are only counting a history if, for some $(x,y) \in S(R)$, the players use that particular communication pattern.

The following two theorems can be proven along the lines of theorems 3.1.1 and 3.2.1:

Theorem 3.3.1 *For every function $f : \{0,1\}^n \to \{0,1\}$ we have*

$$L(f) = \Gamma(R_f)$$

Theorem 3.3.2 *For every monotone function f we have*

$$L_m(f) = \Gamma(R_f^1) = \Gamma(R_f^m)$$

We can now reformulate and generalize fact 2.1.3:

Fact 3.3.1 $C(R) = \Theta(\log \Gamma(R))$ *for every $R \subseteq X \times Y \times Z$.*

It is worthwhile to point out that $\Gamma(R)$ is a more refined complexity measure than $C(R)$.

3.4 Synchronized Protocols

When proving lower bounds for the communication game, it may be convenient to have more structure on the way the players behave. We would like to synchronize the protocol so that the players communicate in rounds. To this end we have the following definition:

Definition 3.4.1 *A (k, l)-protocol is one where the players communicate in rounds in each of which player I sends k bits while player II sends l bits.*

For a relation R, let $r^{k,l}(R)$ be the minimum number of rounds of a (k, l)-protocol for R. The significance of the above definition is expressed in the following theorem:

Theorem 3.4.1 *For every rectangular relation $R \subseteq X \times Y \times Z$, and every integer a,*

1. $r^{a,2^a}(R) \leq C(R)/a$
2. $r^{2^a,a}(R) \leq C(R)/a$

Proof: Let D be a protocol achieving $C(R)$. Divide D into stages of a bits each. A typical stage is associated with subsets $\bar{X} \subseteq X$ and $\bar{Y} \subseteq Y$ which the players partition, in some way, using in all a bits. The idea is to construct a $(a, 2^a)$-protocol which simulates each stage with one round.

For each $x \in \bar{X}$, define a vector $h(x) \in \{0,1\}^{2^a}$ indexed by all possible values of the bits to be sent in the present stage. For a set γ of such values, let $h(x)_\gamma = 1$ iff there is a $y \in \bar{Y}$ for which the players will send values γ had they gotten x, y as inputs. Similarly, define $h(y)$ for every $y \in \bar{Y}$.

The crucial observation is that, because D is deterministic, for each $x \in \bar{X}$ and $y \in \bar{Y}$, there exist a unique γ such that both $h(x)_\gamma = 1$ and $h(y)_\gamma = 1$. To prove 1 (2), the protocol is as follows: player I (II) sends $h(x)$ ($h(y)$) and player II (I) responds with that γ for which $h(x)_\gamma = h(y)_\gamma = 1$. The theorem follows. ∎

As particular cases of theorem 3.4.1 we have the following two corollaries:

Corollary 3.4.1 *For every Boolean function f we have*

1. $r^{a,2^a}(R_f) \leq d(f)/a$
2. $r^{2^a,a}(R_f) \leq d(f)/a$

Corollary 3.4.2 *For every monotone Boolean function f we have*

1. $r^{a,2^a}(R_f^m) \leq d_m(f)/a$
2. $r^{2^a,a}(R_f^m) \leq d_m(f)/a$

3.5 Consequences

The results of § 3 in general, and theorem 3.1.1 in particular, can be used to give alternative definitions to some depth oriented complexity classes such as AC^0 and NC^1. Let us state these definitions explicitly.

Definition 3.5.1 *NC^1 is the set of all Boolean functions f such that*

$$C(R_f) = O(\log n)$$

Definition 3.5.2 *AC^0 is the set of all Boolean functions f such that*

$$r^{\log n, \log n}(R_f) = O(1)$$

In § 4, we will try to demonstrate that the above definitions, and the approach altogether, are quite intuitive. In the meantime, let us make some observations.

Computation, as a whole, can be viewed in many different ways. To state two of them, we can view a computation model either as an *accepting* device (accepting the words of a language, e.g. Turing Machines), or as a *generating* device (generating the words of a language, e.g. Grammars). Here we view a circuit (protocol) as a *separating* device (it separates the words of a language from the non-words). Though this approach is by no means a new one, theorem 3.1.1 presents us with a natural model to pursue it.

Let us take a close look at the meaning of separating $f^{-1}(1)$ from $f^{-1}(0)$. One would like to capture the intuition that a function is complex if its 1's and 0's are hard to distinguish. If one wants to distinguish between two vectors $x, y \in \{0,1\}^n$ knowing only that $x \neq y$ (see definition 4.3.1), one cannot expect to succeed with few bits. The reason is the complete disorganization of the support of the relation in consideration: almost every pair in $\{0,1\}^n \times \{0,1\}^n$ is possible as an input pair. In the other side, when $f^{-1}(1)$ and $f^{-1}(0)$ are organized to form two structured sets, the players will be able to use this organization to direct their search, zooming fast into the answer. The extreme case is when player I always gets an input x such that $x_1 = 1$ while player II gets y with $y_1 = 0$. The structure is so strong that the players don't have to send any messages to distinguish between their vectors. Looking at computation this way may lead us to understand where the heart of complexity lies, and suggests the following meta-idea: *A function f is more complex as $f^{-1}(1)$ and $f^{-1}(0)$ are less organized.* Quantifying this idea into a theorem still seems beyond our present capabilities.

Another advantage of our approach is that it allows us to define the *worst case* complexity or the *average case* complexity of a given protocol. For any relation $R \subseteq X \times Y \times Z$ and any protocol D for it, we can point, in an obvious manner, to the most difficult pair $(x, y) \in S(R)$ for D. It is worth noting that it is not clear how to make a similar definition in the circuit model. The main idea is not to look at the behavior of a circuit on a given input, but to compare it with the behavior of the circuit on those inputs which give different answers.

Size-depth *tradeoffs* for Boolean circuit have been considered, specially for the case of unbounded fanin circuits [Aj83, FSS84, Y85, H86]. Definition 3.5.2 suggests that, perhaps, a cleaner tradeoff to consider is that of number of rounds (alternations) versus length of messages (fanin). See, for example, theorem 6.0.4 in § 6.

Last but not least, the communication game makes transparent the fact that a function f, its *dual*[†] f^*, its *negation* $\bar{f}$, and the negation of its dual $\bar{f}^*$ all have the same depth complexity. Indeed, to get a protocol for R_{f^*} out from a protocol for R_f, one just has to switch the role of the players; to get $\bar{f}$ one just interprets the answer of the protocol accordingly.

† $f^*(x_1, ..., x_n) = \bar{f}(\bar{x}_1, ..., \bar{x}_n)$.

Chapter 4

Miscellaneous Applications

In this chapter we will give some applications of theorem 3.1.1. We will try to demonstrate its usefulness by giving some new intuitive proofs to some old results, and by proving some new ones. The chapter is organized as follows: In § 4.1, we prove, within our context, some results concerning monotone circuits and slices; in § 4.2, we give some examples of protocols for some Boolean functions; in § 4.3, we give both deterministic and randomized bounds for universal relations; finally, in § 4.4, we present a new proof of a theorem of Khrapchenko.

4.1 On Monotone Circuits and Slices

As the first application of the communication complexity approach, we present two theorems which compare monotone with non-monotone circuits. Although both theorems have easy proofs, they both need some careful arguments to formalize them. As absurd as it may seem, the proofs presented here avoid any kind of such technical arguments.

Let us start with some definitions and known facts: For $x \in \{0,1\}^n$, the *weight* of x, $w(x)$, is defined as the number of 1's it contains, i.e. $w(x) = |\{i : x_i = 1\}|$. Let th_k^n be a Boolean function over $\{0,1\}^n$ where for $x \in \{0,1\}^n$, $th_k^n(x) = 1$ iff $w(x) \geq k$. For obvious reasons, we call th_k^n a *threshold* function. The following surprising theorem due to [AKS83] and [V84] will be fundamental in what follows:

Theorem 4.1.1 *$C(R^m[th_k^n]) = O(\log n)$ for every $k = 0, ..., n$.*

It is worth pointing out that both proofs are nontrivial. The proof of [AKS83] has a huge coefficient while the circuits presented in [V84] are non-uniform. Both proofs use a bottom-up approach. As it is pointed out in § 6, our model is best suited for top-down arguments. In fact, one of the main failures of the communication complexity approach has been our inability to understand theorem 4.1.1.

Let f be a Boolean function over $\{0,1\}^n$ and let $0 \leq k \leq n$. The *slice* k of f, f_k, is a Boolean function over $\{0,1\}^n$ such that for $x \in \{0,1\}^n$, $f_k(x) = 0$ if $w(x) < k$; $f_k(x) = f(x)$ if $w(x) = k$; and $f_k(x) = 1$ if $w(x) > k$. The first result we present is due to Berkowitz [B82] and shows that the monotone and non-monotone depth complexities of slice functions are very close to each other.

Theorem 4.1.2 *Let $B_1, B_0 \subseteq \{0,1\}^n$ be such that $B_1 \cap B_0 = \emptyset$ and for every $x \in B_1 \cup B_0$, $w(x) = k$. Let $R(B_1, B_0)$ be as in § 3.1 and let $R^1(B_1, B_0)$ be a relation such that $(x, y, i) \in R^1(B_1, B_0)$ iff $x_i > y_i$. Then,*

$$C(R^1(B_1, B_0)) \leq C(R(B_1, B_0)) + O(\log n).$$

Proof: We construct a protocol D^1 for $R^1(B_1, B_0)$ out of a protocol D for $R(B_1, B_0)$. Let D^1, on (x, y), follow D on (x, y) until it gives an answer i. If $x_i > y_i$ then D^1 terminates. Otherwise, player I thinks of x as if it had $x_i = 1$ so that $k + 1 = w(x) > w(y) = k$. The players then follow the corresponding protocol for $R^m[th^n_{k+1}]$. ∎

Corollary 4.1.1 (Berkowitz) *Let f_k be a slice function over $\{0,1\}^n$. Then*

$$C(R_{f_k}) \leq C(R^m_{f_k}) \leq C(R_{f_k}) + O(\log n).$$

Proof: Follows directly from theorem 4.1.2. ∎

The second result we present is due to Dunne [D84], and shows how to transform a non-monotone circuit into a monotone one by substituting negated variables by certain subcircuits. For a Boolean function f over $\{0,1\}^n$, $i \in [n]$ and $a \in \{0,1\}$, let $f_{i=a} = f(x_1, ..., x_{i-1}, a, x_{i+1}, ..., x_n)$.

Theorem 4.1.3 (Dunne) *Let f be a monotone function, C be a circuit for it and let g be any Boolean function such that, for some i, $f_{i=0} \leq g \leq f_{i=1}$. If we substitute, in C, any appearance of $\bar{x}_i$ by a subcircuit computing g, then the resulting circuit computes f.*

Proof: Note that, as f is monotone, for every $(x, y) \in f^{-1}(1) \times f^{-1}(0)$, there is always an index j such that $x_j > y_j$. Let the players, on (x, y), follow the original protocol for R_f. Suppose that the answer is i and that $x_i = 0$ while $y_i = 1$ (i.e., the players reach a leaf labeled by $\bar{x}_i$). Note that $f_{i=0}(x) = 1$ and $f_{i=1}(y) = 0$ so that $g(x) = 1$ and $g(y) = 0$. The players can now follow a protocol for R_g. ∎

4.2 Upper Bounds

The communication complexity approach to circuit depth is best appreciated, perhaps, by showing protocols for certain Boolean functions. Even though theorem 3.1.1 is constructive, i.e., we can easily construct the protocol associated with a circuit and vice versa, our approach helps us view old circuits through new eyes giving us, in some cases, a better understanding of the problem in hand, if not the possibility of coming out with an improved solution. Alas! Let us look at some examples.

Example 4.2.1 Match: *Given a bipartite graph $G = (X \cup Y, E)$ with $|X| = |Y| = n$, test whether G has a perfect matching.*

Though **Match** has nonuniform NC^2 circuits [KUW87],[MVV87], it is conjectured to require depth $\Omega(n)$ in monotone circuits. In fact, Shamir and Snir [ShS] showed that to compute the related matching polynomial, monotone arithmetic circuits require depth $2n - \log^2 n + \Omega(\log n)$. This bound is achieved (up to a logarithmic additive factor) by the so called Laplace expansion for permanents. For monotone Boolean circuits (as far as we know) this is the best upper bound known.

We will now give a $n + O(\log n)$ bit protocol for R^m[**Match**] improving upon the previous bound. Let us look at the minterms and maxterms of **Match**. Minterms are in one to one correspondence with the bijective mappings $\pi : X \mapsto Y$. Maxterms correspond to $A \times B$ where $A \subseteq X$, $B \subseteq Y$ and $|A| + |B| = n + 1$. A protocol for R^m[**Match**] goes as follows: II sends I the set A (using n bits). Player I looks at $\pi(A)$ as a minterm of the function $th^n_{|A|}$ while player II looks at B as the maxterm of it. The players then follow the corresponding protocol for $R^m[th^n_{|A|}]$. Let $y \in \pi(A) \cap B$, then $(\pi^{-1}(y), y) \in \pi \cap (A \times B)$. We have proved

Theorem 4.2.1 d_m(**Match**) $= C(R^m$[**Match**]$) \leq n + O(\log n)$.

We challenge the reader to understand the above protocol in the circuit model. We feel comfortable enough to state the following conjecture:

Conjecture 1 d_m(**Match**) $= C(R^m$[**Match**]$) = n + \Theta(\log n)$.

As a second example, we present a protocol for the monotone game for the function st-connectivity. In § 5.1, we will give a tight lower bound for this function. We believe that the protocol is much more intuitive than the corresponding circuit.

Example 4.2.2 stconn: *Given the adjacency matrix of an undirected graph $G = (V, E)$ with two distinguished nodes s and t, test whether G has a path from s to t or not.*

Circuits achieving depth $O(\log^2 n)$ can be constructed by raising the adjacency matrix of G to the nth power by recursive squaring. We now present the corresponding protocol for R^m[**stconn**]. Let us view first the minterms and maxterms of **stconn**. Minterms correspond to *st*-paths. Maxterms correspond to *st*-cuts, i.e. to $S \times T$ for some partition of V into S and T where $s \in S$ and $t \in T$. In what follows, we will look at *st*-cuts as colorings $q : V \mapsto \{0, 1\}$ where $q(v) = 0$ iff $v \in S$. Note that an edge in the intersection between an *st*-path and an *st*-cut is just a bichromatic edge in the path.

Intuitively, the players perform a binary search for the bichromatic edge in the path of player I. The protocol consists of $\log n$ stages. Each stage divides the path of player I in half. In each stage, the current path always contains a bichromatic edge and, thus, when the path is a single edge, the protocol terminates. The protocol in each stage is as follows: Player I sends the name of the middle vertex v in the current path p using $\log n$ bits. Player II responds with the color of v. The current path for the next stage is the portion of p where a bichromatic edge is ensured to exist.

There are $\log n$ stages in each of which player I sends $\log n$ bits and player II sends one. We have proved:

Theorem 4.2.2 $d_m(\textbf{stconn}) = C(R^m[\textbf{stconn}]) \leq \log^2 n + \log n.$

Such a natural protocol leads us to the following conjecture:

Conjecture 2 $d_m(\textbf{stconn}) = C(R^m[\textbf{stconn}]) = \log^2 n + \log n.$

4.3 Universal Protocols

We now present an upper bound for all Boolean functions due to Spira [S71]. Though his proof is simple already, we feel that in our language it gains a few points. The upper bound is universal, i.e., it does not depend on the function itself.

In a similar fashion, we define universal relations:

Definition 4.3.1 *Let* $\mathbf{Univ}_n \subseteq \{0,1\}^n \times \{0,1\}^n \times [n]$ *be a relation such that, for every* $x, y \in \{0,1\}^n$ *and* $i \in [n]$, $(x, y, i) \in \mathbf{Univ}_n$ *iff* $x_i \neq y_i$.

We call $\mathbf{Univ}_n$ a universal relation over $\{0,1\}^n$. Note that $S(\mathbf{Univ}_n) = \{(x,y) | x \neq y\}$. For any Boolean function $f : \{0,1\}^n \mapsto \{0,1\}$, we have $R_f = \mathbf{Univ}_n|_{(f^{-1}(1) \times f^{-1}(0))}$ so that $C(R_f) \leq C(\mathbf{Univ}_n)$.

Theorem 4.3.1 $C(\mathbf{Univ}_n) \leq n + \log^* n$.

Proof: Consider the following protocol D_n for $\mathbf{Univ}_n$.

1. I sends $x_1, ..., x_{n-\log n}$.

2. II sends 1 followed by i if $x_i \neq y_i$ for some $i \leq n - \log n$.

   II sends 0 if $x_1, ..., x_{n-\log n} = y_1, ..., y_{n-\log n}$. The players then follow $D_{\log n}$ on $x_{n-\log n+1}, ..., x_n$ and $y_{n-\log n+1}, ..., y_n$ which can be considered as a pair in $S(\mathbf{Univ}_{\log n})$.

It is easy to see that the worst case is when $x_1, ..., x_{n-1} = y_1, ..., y_{n-1}$. In this case, I sends n bits in all while *II* sends $\log^* n$ bits. ∎

By standard counting arguments, we know that the above bound is nearly optimal. The following proof is more direct:

Theorem 4.3.2 $ID_n \leq_2 \overline{\mathbf{Univ}_n}$.

Proof: Recall that any protocol for $\overline{\mathbf{Univ}}_n$ gives the *correct* answer for inputs (x, y) such that $x \neq y$. That is, the protocol provides an i such that $x_i \neq y_i$. With this in mind, we define $\phi_I(x) = x$ and $\phi_{II}(y) = y$ and we note that, for every $i \in [n]$, $I(i) = \{(x, y) : x_i \neq y_i \text{ or } x = y\}$ (recall definition 2.2.11). After running a protocol for $\overline{\mathbf{Univ}}_n$, the players can exchange x_i and y_i and decide whether $x = y$ or not. ▮

Corollary 4.3.1 $C(\mathbf{Univ}_n) = C(\overline{\mathbf{Univ}}_n) \geq C(ID_n) - 2 = n - 1$.

In view of this, and the fact that randomization is usually subsumed by nonuniformity, it is surprising that there exists randomized protocols for $\mathbf{Univ}_n$ which perform much better on the average.

Theorem 4.3.3 *i) LV-C*$(\mathbf{Univ}_n) = O(\log n)$; *ii) MC-C*$(\mathbf{Univ}_n) = O(\log n)$.

Proof: Given a linear function $h(x) = \oplus_{i \in I} x_i$, there is a simple protocol using at most $2 \log |I|$ bits for the relation R_h. The protocol consists of $\log |I|$ stages each of which is associated with a subset $I' \subseteq I$ such that $\oplus_{i \in I'} x_i \neq \oplus_{i \in I'} y_i$. In each stage, the current subset is partitioned into equal sets and, by exchanging the parity of x and y of one of the subsets, the players decide which subset to take as the new current one.

With this in mind, the randomized protocols for $\mathbf{Univ}_n$ are as follows:

Let $h_1, ..., h_{\log n}$ be random linear functions. Recall that the players share the random source so that both know these functions. Let (x, y) be the input.

1. Player I sends $h_i(x)$ for $i = 1 ... \log n$.

2. Player II responds with the an index i such that $h_i(x) \neq h_i(y)$ or 0 if no such index exists.

The players then follow a protocol for R_{h_i} unless player II responds with 0, in which case the Las-Vegas protocol and the Monte-Carlo protocol diverge. The crucial observation is that, for any (x, y) this happens with probability $1/n$.

In the Monte-Carlo protocol the players decide upon an arbitrary answer. In the Las-Vegas protocol, player I sends x to player II who decides upon the answer. We have

$$\begin{aligned} \tilde{D}(x,y) &= (1-\frac{1}{n})O(\log n) + \frac{1}{n}(n + \log n) \\ &= O(\log n) \end{aligned}$$

∎

We now consider a relation which is universal for the monotone game:

Definition 4.3.2 *Let* $\mathbf{Univ}_n^m \subseteq \mathcal{P}([n]) \times \mathcal{P}([n]) \times [n]$ *be a relation such that for every* $x, y \in \mathcal{P}([n])$ *and* $i \in [n]$, $(x, y, i) \in \mathbf{Univ}_n^m$ *iff* $i \in x \cap y$.

Note that $S(\mathbf{Univ}_n^m) = \{(x,y) | x \cap y \neq \emptyset\}$. Again, for any monotone function $f : \{0,1\}^n \mapsto \{0,1\}$, we have $R_f^m = \mathbf{Univ}_n^m|_{(min(f) \times Max(f))}$ so that $C(R_f^m) \leq C(\mathbf{Univ}_n^m)$. A similar protocol to that presented in the proof of theorem 4.3.1 shows:

Theorem 4.3.4 $C(\mathbf{Univ}_n^m) \leq n + \log^* n$.

Consider the following problem:

Example 4.3.1 *Let* $I_n : \mathcal{P}([n]) \times \mathcal{P}([n]) \mapsto \{0,1\}$ *be such that for* $x, y \in \mathcal{P}([n])$, $I_n(x,y) = 1$ *iff* $x \cap y = \emptyset$.

The following theorem can be proved along the lines of theorem 4.3.2:

Theorem 4.3.5 $I_n \leq_2 \overline{\mathbf{Univ}_n^m}$.

So far so good. However, when we deal with randomized complexity, there seems to be a striking difference between the randomized complexities of $\mathbf{Univ}_n$ and $\mathbf{Univ}_n^m$. The reason, perhaps, is that although it is easy to separate two different vectors via a random function, it is not clear how to separate a subset x from subsets y such that $x \cap y = \emptyset$ without separating it from subsets y such that $x \cap y \neq \emptyset$. In fact, we have the following theorem:

Theorem 4.3.6 $MC - C(\mathbf{Univ}_n^m) = \Omega(n)$.

Proof: It is known that $MC - C(I_n) = \Omega(n)$ [KS87].

Theorem 4.3.6 gives evidence about the truth of the fact that non-monotone depth is stronger than its monotone version.

4.4 Khrapchenko's bound

As a nice application of theorem 3.1.1, we give a simple proof of a depth analogue of a theorem of Khrapchenko [K71]. Let C_n be the graph of the n-cube with vertex set $\{0,1\}^n$ and two nodes adjacent iff they differ in exactly one coordinate. Any subset A of edges induces a graph G_A of C_n in the natural way. For a graph G_A and a node x, we denote by $d_A(x)$ and $N_A(x)$ the degree of x in A and the set of neighbors of x in A respectively. We drop the subindex A if no confusion arises.

Theorem 4.4.1 (Khrapchenko) *Let $B_0, B_1 \subseteq \{0,1\}^n$ be such that $B_0 \cap B_1 = \emptyset$. Let $A = C_n \cap (B_0 \times B_1)$. Then, for every function f with $B_0 \subseteq f^{-1}(0)$ and*

$B_1 \subseteq f^{-1}(1)$ *we have*

$$d(f) \geq \log \frac{|A|^2}{|B_0||B_1|}$$

Proof: Consider the relation $R(B_1, B_0)$ as described in § 3.1 and $R' = R(B_1, B_0)|_A$. Let λ be the uniform distribution over A. We will prove that

$$C_\lambda(R') \geq \log \frac{|A|^2}{|B_0||B_1|} \qquad (*)$$

This will prove the theorem as $d(f) = C(R_f) \geq C(R') \geq C_\lambda(R')$. We view $(*)$ as follows: Write

$$\log \frac{|A|^2}{|B_0||B_1|} = \log \frac{|A|}{|B_0|} + \log \frac{|A|}{|B_1|}$$

and notice that $|A|/|B_0|$ and $|A|/|B_1|$ are the average degrees of nodes in B_0 and B_1 respectively. The original proof of this theorem gives no clue whatsoever to the fact that its truth stems from a simple information theoretic fact: *One needs* $\log d$ *bits to distinguish among* d *possibilities.* In what follows, we will claim that the number of bits player I sends is at least the logarithm of the average degree of nodes in B_0 (similarly with player II). Intuitively, this is so because even if player I knows y, he needs $\log d(y)$ bits to tell II which x he has.

We now proceed formally. Fix a protocol D for R. For $(x, y) \in A$, let $b_{\mathrm{I}}(x, y)$ and $b_{\mathrm{II}}(x, y)$ be the number of bits players I and II send when the input to the protocol is (x, y). We have

$$\begin{aligned} D_\lambda &= \frac{1}{|A|}\left[\sum_{(x,y)\in A} (b_{\mathrm{I}}(x,y) + b_{\mathrm{II}}(x,y))\right] \\ &= \frac{1}{|A|}\left[\sum_{x\in B_1}\sum_{y\in N(x)} b_{\mathrm{II}}(x,y) + \sum_{y\in B_0}\sum_{x\in N(y)} b_{\mathrm{I}}(x,y)\right] \end{aligned}$$

We claim:

- For any $x \in B_1$, $\sum_{y \in N(x)} b_{\mathrm{II}}(x,y) \geq d(x) \log d(x)$. This is so because, even if II knows x, he has to tell I which y, among the $d(x)$ possibilities, he has. Formally, this is true for any binary prefix-free encoding of $d(x)$ elements.
- Similarly, for any $y \in B_0$ we have $\sum_{x \in N(y)} b_{\mathrm{I}}(x,y) \geq d(y) \log d(y)$.

We now conclude,

$$\begin{aligned} D_\lambda &\geq \frac{1}{|A|}\left[\sum_{x \in B_1} d(x) \log d(x) + \sum_{y \in B_0} d(y) \log d(y)\right] \\ &\geq \frac{1}{|A|}\left[\sum_{x \in B_1} \frac{|A|}{|B_1|} \log \frac{|A|}{|B_1|} + \sum_{y \in B_0} \frac{|A|}{|B_0|} \log \frac{|A|}{|B_0|}\right] \\ &= \log \frac{|A|^2}{|B_1||B_0|} \end{aligned}$$

where the last inequality follows from the convexity of $x \log x$. ∎

By the definition of A, R' may be considered as a partial function on $B_1 \times B_0$. As noted in § 2.2, theorem 2.2.2 gives a method for proving lower bounds on the complexity of a function. In fact, the original proof of Khrapchenko uses implicitly this method.

Chapter 5

Monotone Depth Lower Bounds

In this chapter we demonstrate the usefulness of the new approach by presenting two monotone depth lower bounds. In § 5.1 we present the depth lower bound for *st*-connectivity. In § 5.2, we present a recent result of Razborov [Ra88] which uses communication complexity to give a monotone lower bound for *MINIMUM COVER.*

5.1 A Lower Bound for st-Connectivity

In this section we give a $\Omega(\log^2 n)$ depth lower bound for monotone circuits computing **stconn**. This section is organized as follows: In § 5.1.1 we give some intuition and we state the main theorem of the section. In § 5.1.2 we give some definitions and useful lemmas. Finally, in § 5.1.3 we give the proof of the theorem.

5.1.1 Intuition

Recall the protocol for $R^m[\mathbf{stconn}]$ presented in section 4.2. The protocol has $\log n$ rounds in each of which player I sends $\log n$ bits while player II sends just one. The crucial observation is that, even if player II would be allowed to send $O(n^\epsilon)$ bits each round (instead of one bit as in the protocol), the players will still need many rounds. Basically, this is so because II doesn't know much about the nodes in I's path. If he sends $O(n^\epsilon)$ bits and the path is of length $O(n^\epsilon)$ then the probability that I gets valuable information from II is negligible. If we could prove a $\Omega(\log n)$ lower bound for the number of rounds needed, we will be able to use corollary 3.4.2 to get the promised $\Omega(\log^2 n)$ lower bound for $C(R^m[\mathbf{stconn}])$.

Note the asymmetry between players I and II. Indeed, if the roles of both players were switched so that player I would be the one who sends $O(n^\epsilon)$ bits per round, they would be able to solve the problem in a constant number of rounds. This is consistent with the intuition obtained by Shamir and Snir in [ShS80].

Let Π_l^n be the set of simple paths on $[n]$ of length l. Define $stconn(l) = R^m[\mathbf{stconn}]|_{(\Pi_l^n \times Max(\mathbf{stconn}))}$ as the restriction of $R^m[\mathbf{stconn}]$ to the case where player I gets only paths of length l. We state the main theorem of this section:

Theorem 5.1.1 *Suppose $l \leq n^{1/10}$. There exists an $0 < \epsilon < 1/2$ such that $r^{\epsilon \log n, n^\epsilon}(stconn(l)) \geq \log l$. In fact, $\epsilon = 1/10$ suffices.*

Corollary 5.1.1 $d_m(\mathbf{stconn}) = \Omega(\log^2 n)$.

Proof: Follows from theorem 5.1.1 and corollary 3.4.2 by taking $l = n^{1/10}$. ∎

Corollary 5.1.2 $L_m(\mathbf{stconn}) = n^{\Omega(\log n)}$.

Proof: Follows directly using fact 2.1.3. ∎

Theorem 5.1.1 is a consequence of the following theorem. Let $vector(l)$ be the game where player I gets an $(l+2)$-vector p with $p[0] = s$ and $p[l+1] = t$ and other entries from $[n]$ and player II gets a coloring $q \in \{0,1\}^n$ of $[n]$ extended so that $q(s) = 0$ and $q(t) = 1$. The goal of the players is to find an element $v \in [n]$ such that for some index i, $p[i] = v$ and either $q(p[i]) \neq q(p[i+1])$ or $q(p[i-1]) \neq q(p[i])$.

Theorem 5.1.2 *Suppose $l \leq n^{1/10}$. There exist an $0 < \epsilon < 1/2$ such that if D is a k-round $(\epsilon \log n, n^\epsilon)$-protocol that solves the game $vector(l)$ for a fraction $\frac{1}{8}n^{-\epsilon}$ of the possible vectors, then $k \geq \log l$.*

Note that in *stconn*(l) the players are asked to find a bichromatic edge while in *vector*(l) they are asked to find an endpoint of a bichromatic edge. Also, in *stconn*(l), player I receives a simple path which can be viewed as a vector without repetitions. Given this, theorem 5.1.1 follows by noting that, by our choice of l, a protocol for *stconn*(l) solves the game *vector*(l) for a fraction $1-o(1)$ of the vectors.

To prove theorem 5.1.2 we will assume, for contradiction, the existence of a k-round $(\epsilon \log n, n^{\epsilon})$-protocol ($k < \log l$) good for a large family of all possible vectors and a large family of all possible colorings. We will pick a large subset of the vectors and colorings for which players I and II sent the same message in the first round. We will give some extra information (by applying a random restriction to the coloring of the nodes) to both players so as to get smaller, yet nicer, subsets which are in 1-1 correspondence with a family of vectors shorter in length (but of higher *quality*) and a family of colorings of fewer nodes. The fact that the original protocol had $(k-1)$ rounds to go, will allow us to find a $(k-1)$-round $(\epsilon \log n, n^{\epsilon})$-protocol for the smaller families. Repeating this k times will give us a protocol without communication that solves a problem which cannot be solved without any messages.

Note the top-down structure of the proof; essentially, the argument shows that, if the output of a shallow circuit, depending on a set of wires, is in some sense complex, then there is a wire which computes a complex subfunction. This is similar to the argument used in [KPPY].

5.1.2 Notation and Definitions

Let $[n]^l$ denote the set of all l-vectors with entries from $[n]$. An *interval* $I \subseteq [l]$ is a subset of consecutive integers. For a vector $p \in [n]^l$ and an interval $I \subseteq [l]$, p_I is the subvector of p in the interval I. For $P \subseteq [n]^l$, $P_I = \{p_I \ : \ p \in P\}$ is the *projection* of P into I. Note that $P_I \subseteq [n]^{|I|}$. Conversely, for $p \in [n]^{|I|}$, $P \subseteq [n]^l$

and an interval I, let $Ext_{P,I}(p) = \{\tilde{p} \in P : \tilde{p}_I = p\}$ be the set of *extensions* of p in P within I. We will drop the subindices P and I if no confusion arises. For $p \in [n]^l$, the *support* of p, $supp(p)$, is defined as the set of nodes contained in p. When no confusion arises, we will denote $supp(p)$ by p. Given a partition of $[l]$ into two intervals L and R, we will denote a vector $p \in [n]^l$ by (p_L, p_R) where each entry is the projection of p into the respective interval.

Similarly, for a coloring $q \in \{0,1\}^n$ and a subset $T \subseteq [n]$, q_T is the *projection* of q into T and for $Q \subseteq \{0,1\}^n$, $Q_T = \{q_T : q \in Q\}$ is the projection of Q into T. For $q \in \{0,1\}^{|T|}$, $Q \subseteq \{0,1\}^n$ and a subset $T \subseteq [n]$, let $Ext_{Q,T}(q) = \{\tilde{q} \in Q : \tilde{q}_T = q\}$ be the set of *extensions* of q in Q within T (again, we drop the subindices Q and T whenever possible). For a *restriction* $\rho : [n] \mapsto \{0,1,*\}$, we will denote by Q^ρ the set of colorings in Q consistent with ρ, (i.e., $\{q \in Q : \rho(i) \neq * \Rightarrow \rho(i) = q_i\}$).

For a subset A of a universe Ω, the *density* of A, $\mu_\Omega(A)$, is defined as $|A|/|\Omega|$. In what follows, we will work with densities rather than with cardinalities. If Ω is clear from the context, we will drop the subindex and write $\mu(A)$. The reader should be aware that we may mix densities with respect to different universes in the same equation.

We will need the following combinatorial lemma due to J. Hastad and R. Boppana: Let $H \subseteq A_1 \times ... \times A_k$ and for $v \in A_i$, let $Ext_{A_i}(v) = \{u \in H : u_i = v\}$. Note that, though $Ext_{A_i}(v) \subseteq H$, $Ext_{A_i}(v)$ will be considered as a subset of $H/A_i = A_1 \times ... \times A_{i-1} \times A_{i+1} \times ... \times A_k$ and, in what follows, its density will be defined with respect to H/A_i.

Lemma 5.1.1 *Let* $H \subseteq A_1 \times ... \times A_k$. *Let* $B_i = \{u \in A_i : \mu(Ext_{A_i}(u)) \geq \mu(H)/2k\}$. *Then*

$$\prod_{i=1}^{k} \mu(B_i) \geq \frac{\mu(H)}{2}$$

Proof: Say that a member $(u_1, ..., u_k)$ of H is *bad* if for some i, $u_i \notin B_i$. Let $\bar{H}$ be the set of bad elements in H. We have

$$\mu(\bar{H}) \leq \sum_{i=1}^{k} \mu\left(\cup_{u \notin B_i} Ext_{A_i}(u)\right) < \sum_{i=1}^{k} \frac{\mu(H)}{2k} = \frac{\mu(H)}{2}$$

the lemma follows immediately, by noting that

$$\prod_{i=1}^{k} \mu(B_i) \geq \mu(H) - \mu(\bar{H})$$

∎

Corollary 5.1.3 *If $k = 2$, then there exists an i such that $\mu(B_i) \geq \left(\frac{\mu(H)}{2}\right)^{1/2}$.* ∎

Corollary 5.1.4 $\Pr\left(\mu(B_i) < (\mu(H)/2)^{2/k}\right) < 1/2$ *for i chosen randomly from* $\{1, ..., k\}$. ∎

5.1.3 The proof

Proof: [of theorem 5.1.2]

In what follows, we will only consider $(\epsilon \log n, n^{\epsilon})$-protocols. The existence of ϵ will be clear from the proof, though one can check that $\epsilon = 1/10$ suffices. We will define a sequence of problems of different sizes as follows: We first define the parameters of the problems. Let $t_{max} = \log l - 1$.

Let $n_0 = n$ and $n_{t+1} = n_t - 4{n_t}^{1/2}$. Note that

$$n/2 \leq n_t \leq n \qquad \text{for} \qquad t \leq t_{max} \tag{1}$$

Let $l_0 = l$ and $l_{t+1} = l_t/2$ and note that

$$2 \leq l_t \leq l \qquad \text{for} \qquad t \leq t_{max} \tag{2}$$

Consider the following property:

H(t, k): *There exist a collection of vectors* $P_t \subseteq [n_t]^{l_t}$ *of length* l_t*, and a collection of colorings* $Q_t \subseteq \{0,1\}^{n_t}$ *of* $[n_t]$*, with* $\mu(P_t) \geq \frac{1}{8}n^{-\epsilon}$ *and* $\mu(Q_t) \geq 2^{-2tn^{\epsilon}}$ *such that* $r^{\epsilon \log n, n^{\epsilon}}(vector(l_t)|_{(P_t \times Q_t)}) \leq k$*. Let* D_t *be a* $(\epsilon \log n, n^{\epsilon})$*-protocol achieving* $r^{\epsilon \log n, n^{\epsilon}}(vector(l_t)|_{(P_t \times Q_t)})$.

We will prove the following two claims:

Claim 5.1.1 *For* $t \leq t_{max}$ $\neg H(t, 0)$.

Claim 5.1.2 *For* $t \leq t_{max}$ $H(t,k) \rightarrow H(t+1, k-1)$

It is clear that the two claims imply $\neg H(0, t_{max})$ which in turn implies our theorem.

The first claim follows easily by noticing that there is not a single node (other than s and t) which appears in every vector of P_t so that player II cannot know the answer. To see this, note that the fraction of vectors of length l_t which contain a given node is $1 - (1 - 1/n_t)^{l_t} \ll \frac{1}{8}n^{-\epsilon}$. This is enough for proving the claim as both players must know the answer. However, it can also be shown that, for most input pairs, player I will not know the color of a single node in its vector.

The second claim will be proved by assuming $H(t,k)$ and constructing P_{t+1}, Q_{t+1} and D_{t+1} so as to satisfy $H(t+1, k-1)$. Take P_t, Q_t and D_t which satisfy $H(t,k)$. Let us look at the protocol after the first round. By the pigeonhole principle, there exist $P \subseteq P_t$ with $\mu(P) \geq \frac{1}{8}n^{-2\epsilon}$ such that for every vector in P, I sent the same message. Similarly, there exists $Q \subseteq Q_t$ with $\mu(Q) \geq 2^{-(2t+1)n^{\epsilon}}$ so that for every coloring of Q, II sent the same message.

Let $L = \{1, ..., l_t/2\}$ and $R = \{l_t/2 + 1, ..., l_t\}$ be a partition of the vector's coordinates into left and right intervals of the same length. We say that P is L-

good if *many* left projections of P have, each, *many* extensions to the right; that is, if

$$\mu\left(\{p_L : \mu(Ext_{P,L}(p_L)) \geq n^{-2\epsilon}/32\}\right) \geq \frac{1}{4}n^{-\epsilon}$$

R-goodness is defined similarly. The following lemma says that if we shrink the length of the vectors to half and we restrict our family P to one of the intervals, then we can improve the quality of our collection. This is one of our main ideas: Although we cannot raise the absolute size of P, by reducing the size of the universe we can increase its density (quality).

Lemma 5.1.2 *P is either L-good or R-good.*

Proof: The lemma follows using corollary 5.1.3 and viewing P as a subset of $[n_t]^{l_{t+1}} \times [n_t]^{l_{t+1}}$. ∎

Without loss of generality, assume that P is L-good and let A be the set of vectors in P_L with many extensions. The next lemma is the heart of our argument:

Lemma 5.1.3 *There exists a restriction $\rho : [n_t] \mapsto \{0, 1, *\}$ with $|\rho^{-1}(*)| = n_{t+1}$ such that the following properties hold:*

G1: $\mu(Q^{\rho}_{\rho^{-1}(*)}) \geq 2^{-2(t+1)n^{\epsilon}}$.

G2: *$\exists \bar{P} \subseteq P$ such that*

- $\forall p \in \bar{P},\ p_L \subseteq \rho^{-1}(*)$ *and* $p_R \subseteq \rho^{-1}(1)$
- $\forall p, p' \in \bar{P}\ \ p_L \neq p'_L$
- $\mu(\bar{P}_L) \geq \frac{1}{8}n^{-\epsilon}$.

Assuming the lemma is true, we will finish the proof of the second claim:

Let $Q_{t+1} = Q^{\rho}_{\rho^{-1}(*)}$ and $P_{t+1} = \bar{P}_L$ and rename the coordinates so that $[n_{t+1}] = \rho^{-1}(*)$. Note that there is a natural 1-1 correspondence between Q_{t+1} and Q^{ρ} and between P_{t+1} and $\bar{P}$. Also note that for every $q \in Q^{\rho}$ and for every $p \in \bar{P}$ any bichromatic edge lies in the interval L. The protocol D_{t+1} on (P_{t+1}, Q_{t+1}) simulates the remaining rounds of the protocol D_t on $(\bar{P}, Q^{\rho})$ by following the behavior of the associated vector and coloring. ∎

Proof: [of lemma 5.1.3]

In what follows, we denote $V = [n_t]$, $v = n_t$, $l = l_t$, $l' = l_{t+1}$ for simplicity's sake. The existence of a good restriction will be shown by probabilistic methods. We will pick ρ uniformly from the set of all restrictions with $|\rho^{-1}(*)| = v - 4\sqrt{v}$ and $\Pr(\rho(x) = 0 | \rho(x) \neq *) = 1/2$, and show that, with positive probability, the conditions of the lemma are fulfilled. Specifically, we will show that $\Pr(\neg \mathbf{G1}) + \Pr(\neg \mathbf{G2}) \leq 1/2 + o(1)$.

Let us start with **G1**: Intuitively, the following lemma says that, with high probability, ρ does not give player I too much information about the colors of nodes in $\rho^{-1}(*)$.

Lemma 5.1.4 $\Pr\left(\mu(Q^{\rho}_{\rho^{-1}(*)}) < 2^{-2(t+1)n^{\epsilon}}\right) \leq \frac{1}{2} + o(1)$.

Proof:

Let $\alpha = 2^{-(2t+1)n^{\epsilon}}$. Picking ρ uniformly from all restrictions with $|\rho^{-1}(*)| = v - 4\sqrt{v}$ is equivalent to picking randomly $T = \rho^{-1}(1) \cup \rho^{-1}(0)$ among all $4\sqrt{v}$-subsets of V, and then picking the restriction of ρ to T randomly among all vectors x in $\{0,1\}^{4\sqrt{v}}$. Let $k = \sqrt{v}/4$. Say T is *bad* if

$$\mu\left(\left\{x : \mu(Ext_{Q,T}(x)) \geq \frac{\alpha}{2k}\right\}\right) < \left(\frac{\alpha}{2}\right)^{2/k},$$

T is *good* otherwise. We have

$$\Pr\left(\mu(Q^{\rho}_{\rho^{-1}(*)}) < \frac{\alpha}{2k}\right) \leq \Pr(T \text{ is bad}) + \Pr\left(\mu(Q^{\rho}_{\rho^{-1}(*)}) < \frac{\alpha}{2k} \mid T \text{ is good}\right)$$

Note that $Q^{\rho}_{\rho^{-1}(*)} = Ext_{Q,T}(x)$. By the definition of goodness, and the choice of k, the second term is bounded by $1-(\alpha/2)^{2/k} = o(1)$. Also note that $\alpha/2k \geq 2^{-2(t+1)n^{\epsilon}}$. It remains to bound the first term: We pick a random T by first picking a random partition of V into $4\sqrt{v}$-subsets and then picking a random subset from the partition. We must show that for any partition, $\Pr(T \text{ is bad}) < 1/2$ for a random T in the partition. But this is precisely the content of corollary 5.1.4. ∎

Now we take care of **G2**:

Let $A^* = \{p \in A : p \subseteq \rho^{-1}(*)\}$. We say that ρ *kills* a vector $p_L \in A$ if there is no $p_R \in Ext(p_L)$ with $p_R \subseteq \rho^{-1}(1)$. We will show that for every choice of ρ, $\mu(A^*)$ is large and hence $\Pr(\neg\mathbf{G2}) \leq \Pr(\exists p_L \in A \text{ killed by } \rho)$.

Claim 5.1.3 *For every ρ, $\mu(A^*) \geq \frac{1}{8}n^{-\epsilon}$.*

Proof: Recall that $|\rho^{-1}(*)| = v - 4\sqrt{v}$ so that $|V \backslash \rho^{-1}(*)| = 4\sqrt{v}$. It is easy to see that at most a fraction $1-[(v-4\sqrt{v})/v]^{l'} \leq \frac{1}{8}n^{-\epsilon}$ of the vectors in $[v]^{l'}$ intersect $V\backslash\rho^{-1}(*)$ so that at least a fraction $\frac{1}{4}n^{-\epsilon} - \frac{1}{8}n^{-\epsilon} \geq \frac{1}{8}n^{-\epsilon}$ of them are in A^*. ∎

We now bound the probability that there exists a vector in A killed by ρ:

Claim 5.1.4 $\Pr(\exists p_L \in A \quad \text{killed by} \quad \rho) = o(1)$.

Proof: We have

$$\Pr(\exists p_L \in A \text{ killed by } \rho) \leq |A| \cdot \max_{p_L \in A}\{\Pr(p_L \text{ is killed by } \rho)\}$$

so let us look at the worst possible $p_L \in A$. Let $F \equiv (p_L$ is killed by $\rho)$. Note that $\Pr(F)$ depends only on $\rho^{-1}(1)$. We pick $\rho^{-1}(1)$ as follows: Pick a number t between 0 and $4\sqrt{v}$ according to the binomial distribution (i.e. $\Pr(t = i) = \binom{4\sqrt{v}}{i}2^{-4\sqrt{v}}$). If $t < \sqrt{v}$, we assume that F fails. Otherwise, we pick a subset $T = \rho^{-1}(1)$ where $|T| = t$ by choosing $\sqrt{v}/l'$ independent random vectors $y_1 \ldots y_{\sqrt{v}/l'}$ from $[v]^{l'}$, puting all nodes in these vectors in T, and adding enough random nodes so that $|T| = t$. It is clear that we are simulating our original distribution on $\rho^{-1}(1)$. We can now estimate $Pr(F)$ by

$$\begin{aligned}
Pr(F) &\leq \Pr(t < \sqrt{v}) + \Pr(F | t \geq \sqrt{v}) \\
&\leq (2/e)^{\sqrt{v}} + \Pr(\forall i \ y_i \notin Ext(p_L)) \\
&\leq (2/e)^{\sqrt{v}} + (1 - \mu(Ext(p_L)))^{\sqrt{v}/l'} \\
&\leq \exp(-n^{1/5})
\end{aligned}$$

where we are using Chernoff's bound to estimate $\Pr(t < \sqrt{v})$ [Ch].

Recalling that $|A|$ is less than $n^{n^{1/10}}$, we easily conclude our calculations and get

$$\Pr(\exists p \in A \text{ killed by } \rho) \leq n^{n^{1/10}} \cdot \exp(-n^{1/5}) = o(1)$$

∎

We have $\Pr(\neg\mathbf{G1}) + \Pr(\neg\mathbf{G2}) \leq 1/2 + o(1)$ implying the existence of a good restriction. Take any consistent extension of each $p_L \in A^*$ not killed by ρ to form $\bar{P}$. We have $\mu(\bar{P}_L) \geq \frac{1}{8}n^{-\epsilon}$ and lemma 5.1.3 is proved. ∎

5.2 Lower Bounds Via Reductions

In this section we present an informal discussion of recent results of Razborov [Ra88] who uses reductions to prove monotone depth lower bounds. As stated in section 2.2.1, there exist known techniques for proving lower bounds for the communication complexity of relations $R \subseteq X \times Y \times Z$ where for every $x \in X$ and $y \in Y$, there exist a unique $z \in Z$ such that $(x, y, z) \in R$. That is, when R defines a function $r : X \times Y \mapsto Z$. An appealing possibility is the use of reductions as defined in definitions 2.2.9 or 2.2.11 together with lemma 2.2.2 or lemma 2.2.3 to give lower bounds for general relations.

In general, given $f : \{0,1\}^n \mapsto \{0,1\}$ and a function $F : X \times Y \mapsto Z$, the idea is to reduce F to either R_f or R_f^m (if f is monotone) and then use either theorem 2.2.2 or theorem 2.2.3 to give a lower bound for $C(F)$ which, in turn, implies lower bounds for either $C(R_f)$ or $C(R_f^m)$.

5.2.1 Reductions and Monotone Lower Bounds

Consider the following monotone function:

Example 5.2.1 $\mathbf{MC_{k,n}}$: *Given a bipartite graph $G = (U \cup V, E)$ with $|U| = |V| = n$, test whether there exists a $U' \subseteq U$ with $|U'| = k$ such that every node in V has a neighbor in U'.*

The following theorem of Razborov is a concrete realization of the ideas described above:

Theorem 5.2.1 (Razborov) *There exists a constant $c > 0$ such that if $k = c \log n$, then*

$$I_{k,n} \leq_1 R^m[\mathbf{MC_{k,n}}]$$

where $I_{k,n} : [n]^k \times [n]^k \mapsto \{0,1\}$ *is defined by* $I_{k,n}(x,y) = 1$ *iff* $x \cap y = \emptyset$. *(see corollary 2.2.2).*

Proof: We are going to use the following auxiliary lemma:

Lemma 5.2.1 *There exist a* $c > 0$ *and sets* $z_1, ..., z_n \subseteq [n]$ *such that for every* $x, y \in [n]^{c \log n}$ *where* $x \cap y = \emptyset$, *there exists a* z_i *such that* $x \subseteq z_i$ *and* $y \cap z_i = \emptyset$.

Proof: [of lemma] We are going to use a probabilistic construction. A random $z \subseteq [n]$ *separates* a pair $x, y \in [n]^{c \log n}$ such that $x \cap y = \emptyset$ with probability $2^{-2c \log n}$. Therefore, if we pick c small enough, n random subsets will separate *every* pair of subsets with probability greater than zero. The lemma follows. ▮

Intuitively, given x, y such that $x \cap y = \emptyset$, one can think of any z_i such that $x \subseteq z_i$ and $y \cap z_i = \emptyset$ as a *certificate* for the fact that $x \cap y = \emptyset$. Similarly, if $i \in x \cap y$, one can think of i as a certificate for the fact that $x \cap y \neq \emptyset$.

Let $H = (U \cup V, E_H)$, $|U| = |V| = n$, be a bipartite graph such that $(u_i, v_j) \in E_H$ iff $i \in z_j$ where $z_1, ..., z_n$ are as in lemma 5.2.1. The idea is to use H as the *base* graph for our reduction in such a way that any answer for $R^1[\mathbf{MC_{k,n}}]$ will provide us with a certificate for either $x \cap y = \emptyset$ or $x \cap y \neq \emptyset$.

In what follows, let $k = c \log n$ where c is as in lemma 5.2.1. For $x, y \in [n]^k$ consider the following set of edges in $U \times V$.

$$\begin{aligned} A_I(x) &= \{(u_i, v_j) : i \notin x\} \\ B_I(x) &= \{(u_i, v_j) : z_j \cap x = \emptyset\} \\ A_{II}(y) &= \{(u_i, v_j) : i \in y\} \\ B_{II}(y) &= \{(u_i, v_j) : y \not\subseteq z_j\} \end{aligned}$$

We now define the reduction functions. Recall that $\phi_I(x) : [n]^k \mapsto f^{-1}(1)$ and $\phi_{II}(y) : [n]^k \mapsto f^{-1}(0)$.

- $\phi_I(x) = (H \cup B_I(x)) - A_I(x)$
- $\phi_{II}(y) = (H \cup B_{II}(y)) - A_{II}(y)$

We have to show that $\phi_I(x) \in \mathbf{MC_{k,n}}^{-1}(1)$ and $\phi_{II}(y) \in \mathbf{MC_{k,n}}^{-1}(0)$.

1. $\phi_I(x) \in \mathbf{MC_{k,n}}^{-1}(1)$: We have to show that there exists a $U' \subseteq U$, $|U'| = k$, such that every $v \in V$ is covered by a node in U'. Take $U' = \{u_i : i \in x\}$. Edges in $A_I(x)$ only concern nodes not in U'. Any node $v_j \in V$ is either such that $x \cap z_j = \emptyset$ in which case edges from $B_I(x)$ cover it, or is such that $x \cap z_j \neq \emptyset$ in which case, by definition, H contains an edge from U' to it. Also $|U'| = |x| = k$.

2. $\phi_{II}(y) \in \mathbf{MC_{k,n}}^{-1}(0)$: We have to show that for every $U' \subseteq U$, $|U'| = k$, there exists a node $v \in V$ not covered by U'. A first observation is that nodes $u_i \in U'$ such that $i \in y$ cannot contribute because $A_{II}(y)$ contains all the edges emanating from them. Let $w = \{i : u_i \in U'\}$. Without loss of generality $w \cap y = \emptyset$. By the claim, there exists a z_j such that $y \subseteq z_j$ and $z_j \cap w = \emptyset$. This node is not covered by edges from $B_{II}(y)$ and, by definition, neither by edges from H.

Now suppose that $(u_i, v_j) \in \phi_I(x)$ and $(u_i, v_j) \notin \phi_{II}(y)$. That is, (u_i, v_j) is a possible answer for $R^1[\mathbf{MC_{k,n}}]$ on input $(\phi_I(x), \phi_{II}(y))$. The way we constructed $\phi_I(x)$ and $\phi_{II}(y)$ implies that either of the following alternatives holds:

1. $(u_i, v_j) \in A_{II}(y) - A_I(x)$, or

2. $(u_i, v_j) \notin A_{II}(y)$ so that $(u_i, v_j) \in B_I(x) - B_{II}(y)$.

Furthermore, if the players can decide which alternative holds, then they can decide whether $x \cap y = \emptyset$ or not. This is because alternative 1 implies that $i \in x \cap y$, while alternative 2 implies that $x \cap z_j = \emptyset$ and $y \subseteq z_j$.

To finish the protocol for $I_{k,n}$, player II sends a bit saying which of the two alternatives is true, i.e. whether $(u_i, v_j) \in A_{II}(y)$ or not. In the notation of definition 2.2.11, $C(I_{k,n}|_I((u_i, v_j))) \leq 1$. ∎

Corollary 5.2.1 *For $k = O(\log n)$,*

$$d_m(\mathbf{MC_{k,n}}) = C(R^m[\mathbf{MC_{k,n}}]) \geq \log \binom{n}{k} - 1.$$

Proof: Follows from corollary 2.2.2 and lemma 2.2.3. ∎

5.2.2 Reductions and Partial Functions

For any Boolean function f, the relation R_f can be viewed as a search problem: Given $x \in f^{-1}(1)$ and $y \in f^{-1}(0)$, *find* an index i such that $x_i \neq y_i$. The following observation of Razborov reduces partial functions to these search problems allowing us to work with partial functions instead.

A *partial function* $F : X \times Y \mapsto Z$ is a function such that $S(F) \neq X \times Y$. Consider any field K and look at F as a function from $X \times Y$ to K. As in § 2.2.2, we define the matrix M_F over $K \cup \{*\}$ with rows and columns indexed by X and Y respectively and with (x, y)-entry equal to $F(x, y)$ if $(x, y) \in S(F)$ or $*$ if $(x, y) \notin S(F)$. A matrix M over K is an *extension* of M_F if, for every $(x, y) \in S(F)$, the (x, y)-entries of M and M_F are equal. The following theorem is a trivial extension of theorem 2.2.3:

Theorem 5.2.2 $C(F) \geq \min_M \log rk(M)$, *where the minimum is taken over all extensions M of M_F.*

Of course, in order to use theorem 5.2.2, one will have to prove lower bounds for a possible large number of matrices. The following observation of Razborov shows that, in principle, one can use theorem 5.2.2 to get lower bounds for $C(R_f)$ for any function f.

For a Boolean function $f : \{0,1\}^n \mapsto \{0,1\}$ and a partition of $[n]$ into two disjoint sets I and J, let $R_f^{I,J} \subseteq f^{-1}(1) \times f^{-1}(0) \times \{0,1\}$ be such that $(x, y, 0) \in R_f^{I,J}$ only if $x_I \neq y_I$ (i.e., if there is an $i \in I$ such that $x_i \neq y_i$) and $(x, y, 1) \in R_f^{I,J}$ only if $x_J \neq y_J$. Notice that one can think of $R_f^{I,J}$ as a partial function where $(x, y) \in S(R_f^{I,J})$ iff $x_I = y_I$ or $x_J = y_J$. In such cases, the answer to $R_f^{I,J}$ is determined by (x, y).

Intuitively, the answer to R_f consists of $\log n$ bits. For any partition of $[n]$ into I and J, the relation $R_f^{I,J}$ asks for one bit of the answer. It is clear that asking for one bit is much less than asking for $\log n$ bits. Formally:

Lemma 5.2.2 *For every partition of* $[n]$ *into* I *and* J, $R_f^{I,J} \leq R_f$.

It may be, however, that for any function f and for any partition of $[n]$ into I and J, $R_f^{I,J}$ is easy. The following theorem of Razborov [Ra88], though proved by counting arguments, shows not only that this is not the case, but also that one can, in principle, use *rank* arguments to prove lower bounds.

Theorem 5.2.3 (Razborov) *For large enough* n, *and for most functions* $f : \{0,1\}^n \mapsto \{0,1\}$, *the possible bound attainable via theorem 5.2.2 is* $\Omega(n)$.

Chapter 6

Discussion and Future Research

We have presented a new approach to circuit depth. We have demonstrated, by way of examples, that the approach is very intuitive and that it provides us with the correct setting for understanding some old results (see § 4). Furthermore, we have succeeded, using this approach, in proving a nontrivial monotone depth lower bound. In this chapter, we would like to comment on some points regarding the approach and propose some open problems.

- The new model suggests new ways of looking at computation and provides us with some clues as to where to look for the cause of complexity (see § 3.5).
- We presented examples of upper bounds for some Boolean functions. In one case (example 4.2.1) we succeeded in improving the upper bound via a trivial protocol whose associated circuit seems completely counterintuitive. We feel that the communication model is best suited for algorithms which can be presented in a natural top-down fashion.
- Reflecting a bit on the proof of theorem 5.1.1, we see that the lower bound still works for a very small subset of $S(stconn(l))$. In fact, we believe that other lower bounds will have the same flavor, i.e., will work on a small subset of the support of the relation in consideration. The reason is that, as two *close* functions have *similar* complexities, two *similar* relations have *similar* complexities. That is, any lower bound proof must also work for a set of functions close to the one in consideration. In view of theorems 4.3.3 and 2.2.1 we know that, for every function f, there is a deterministic protocol of complexity $O(\log n)$ which gives a correct answer on at least a fraction $1 - 1/n$ of the input pairs in $S(R_f)$. Does this show an inherent limitation in proving lower bounds of the form $\omega(\log n)$?

- The main success of the new model (as of today) is, no doubt, the lower bound for *st*-connectivity. Though the lower bound can now be expressed cleanly in the circuit model (see [BS88]), we feel that we could have never proven the result without the intuition obtained through the new approach. We have failed, however, in presenting a general technique for proving monotone depth lower bounds (see [An85] or [Ra85a] for such techniques for proving monotone size lower bounds). It is striking the fact that it is not clear how to adapt some of the ideas developed in this work to prove lower bounds for other functions (see the open problems below).

- Theorem 3.1.1, which presents us with the equivalence between communication complexity and circuit depth, is very constructive. That is, given a circuit for a function f, the theorem shows how to define a protocol for the relation R_f and vice versa. As a consequence, any argument given in any of the two models can be interpreted in the other model. In fact, many of the results of chapter 4 are nothing more than interpretations, in the model of communication complexity, of old results which were originally presented in the circuit model. There have been also interpretations in the other direction. In particular, in BS[88] there is an interpretation of corollary 5.1.1 in the circuit model without ever mentioning the word communication.

  There are, however, some arguments in each of the models which do not have *natural* counterparts in the other model. In one direction, some of the results obtained via bottom-up arguments in the circuit model do not have natural analogues in the new model (e.g. open 6.0.4 below). In the other direction one can think also on many notions which come about naturally in the communication model but which do not have natural counterparts in the circuit model. In particular, not every relation R is of the form R_f or R_f^m for some function f. In this way, if we show that R can be reduced to a relation associated with a function, and we work with R, we may use arguments that do not have a natural interpretation in the circuit model (e.g. example 6.0.3 below).

We would like now to propose some open problems. We have chosen problems which, we feel, will give clues as to what kind of arguments will have a chance for non-monotone lower bounds.

- The lower bound proof for st-connectivity seems to depend heavely on the fact that player I is allow to send no more than $\epsilon \log n$ bits per round, and the fact that the length of the paths in consideration is relatively small (n^ϵ). We feel that, in order to generalize the ideas presented here, one will have to understand the above limitations. In order to do so, we propose the following open problems:

Open 6.0.1 *Show that a $(k \log n, n^\epsilon)$-protocol for stconn(l) requires $\Omega(\log_{k+1} l)$ rounds. Note that a trivial generalization of the protocol presented in § 4.2 achieves this bound.*

Open 6.0.2 *Show that $C(stconn(n)) = \Omega(\log^2 n)$. Recall that stconn$(n)$ is the restriction of R^m[**stconn**] to the case where player I gets only paths of length n.*

It is, of course, a main challenge to prove non-trivial lower bounds for other monotone functions. In view of problem 6.0.2, a natural function to consider will be:

Example 6.0.2 conn: *Given an undirected graph G, test whether G is connected or not.*

Open 6.0.3 *Show that $C(R^m$[**conn**]$) = \Omega(\log^2 n)$.*

Note that problem 6.0.2 above implies this one.

- In § 5.1, we showed that, to solve *stconn*(l), the players have to perform a *binary* search for the bichromatic edge in the path of player I. After playing a bit with the function **Match** as defined in § 4.2, it seems that the players have to perform a *sequential* search for the edge in $\pi \cap (A \times B)$. This is the reason for our conjecture 1, though it would be nice to prove even that:

  Open 6.0.4 *Show that* $C(R^m[\mathbf{Match}]) = \Omega(n^\epsilon)$.

  Razborov [Ra85b] has shown that $C(R^m[\mathbf{Match}]) = \Omega(\log^2 n)$ by methods which apply to size rather than to depth. A solution to *Open* 6.0.4 and the fact that $C(R[\mathbf{Match}]) = O(\log^2 n)$ ([KUW85], [MVV87]) will settle in the affirmative the question whether negation saves in depth. (This question, for size, has been settled in the affirmative by Razborov [Ra85b] and, in a stronger form, by Tardos [T87]). Theorem 4.3.6 gives evidence about the truth of this question.

- In order to prove lower bounds, one needs a very deep intuition into why a function is complex. One has to be able to recognize the easy functions from the hard ones. In this respect, there are still mysteries which confront our intuition. The most striking one is the monotone relation $R^m[\mathbf{Maj}]$ associated with *majority**.

  Even though our intuition says that $C(R^m[\mathbf{Maj}]) = \Omega(\log^2 n)$, results of Ajtai, Komlós and Szemerédi [AKS83] and Valiant [V84], as stated in § 4.1, show that $C(R^m[\mathbf{Maj}]) = O(\log n)$. We will like to understand this phenomenon in order to get our intuition right:

  Open 6.0.5 *Give a natural protocol achieving* $C(R^m[\mathbf{Maj}]) = O(\log n)$.

---

*$R^m[\mathbf{Maj}] \subset [2n-1]^n \times [2n-1]^n \times [2n-1]$ where for $p, q \in [2n-1]^n$ and $i \in [2n-1]$, $(p, q, i) \in R^m[\mathbf{Maj}]$ iff $i \in p \cap q$.

- The ideas developed in § 5.1.3 do not work for nonmonotone lower bounds. The reason is that we worked directly, and our ideas were tailored for, the monotone game. To see this, we point out that there is a simple and cheap protocol for the restriction of R[**stconn**] to the case where player I gets a graph which consists of a single st-path, and player II gets a graph which consists of a pair of disjoint cliques (which correspond to an st-cut): Player I sends the name of 3 consecutive nodes in its path. It is easy to see that an edge from the triangle defined by the 3 nodes is an answer. We do not see, however, any inherent limitation in using theorem 3.1.1 together with the general approach of looking at computation top-down, in order to get non-monotone lower bounds. To train ourselves, we propose to look at old results and prove them via a top-down argument. In particular, we have the following result which was proved using a bottom-up technique [Y85], [H86]:

Theorem 6.0.4 (Yao, Hastad) *Let $\oplus_n$ be the parity function on n variables. Then $r^{k,k}(R_{\oplus_n}) = \Theta(\log_{k+1} n)$.*

We would like to have a top-down proof of the following very clean problem:

Open 6.0.6 *Show that $r^{k,\log(k+1)}(R_{\oplus_n}) = \log_{k+1} n$.*

- A general outline for a lower bound proof for the communication complexity of a relation R is:

  1. Pick a suitable $I \subseteq S(R)$ which is structured enough to work with.
  2. Prove a lower bound for $C(R|_I)$.
  3. Use lemma 2.2.1 to conclude a lower bound for $C(R)$.

The main problem is that, for most I, $R|_I$ may be considerably easier than R. As a concrete example, consider any Boolean function f together with R_f. If $I \subseteq S(R_f)$ is such that for every $(x, y) \in I$, $w(x) \neq w(y)$, then the players can, by exchanging first $w(x)$ and $w(y)$, distinguish between their vectors using a protocol for a threshold function separating $w(x)$ from $w(y)$ (see § 4.1). That is, $C(R_f|_I) = O(\log n)$ independently of the function f. This implies that any I used on the outline described above must contain only vectors from the same slice [†].

In view of the last point, we would like to propose the following relation as a candidate for separating NC^1 from L[‡]. The relation is related to the problem of deciding whether a given permutation is acyclic. This problem was shown to be complete for L in [CMcK87] so that we can safely assume that **Cycle** is hard.

Example 6.0.3 Cycle: *A subset of* $S_n \times S_n \times [n]$ *where* $(\pi_1, \pi_2) \in S(\mathbf{Cycle})$ *iff* π_1 *is acyclic and* π_2 *contains at least 2 cycles. For* $(\pi_1, \pi_2) \in S(\mathbf{Cycle})$ *and* $i \in [n]$, $(\pi_1, \pi_2, i) \in \mathbf{Cycle}$ *iff* $\pi_1(i) \neq \pi_2(i)$.

The crucial observation is that, if we consider the relation $R' \subseteq R[\mathbf{conn}]|_I$ where $(x, y) \in I$ iff x encodes a graph consisting of a single hamiltonian cycle while y encodes a graph consisting of a disjoint union of at least 2 cycles, then, in the notation of § 2.2,

$$\mathbf{Cycle} \leq R' \leq_{O(\log n)} \mathbf{Cycle}$$

so that

$$C(\mathbf{Cycle}) \leq C(R') \leq C(\mathbf{Cycle}) + O(\log n).$$

Note that, for every $(x, y) \in I$, both x and y have exactly n edges. Moreover, every node in x and y has degree 2. This makes the graphs fairly similar and difficults the use of naive protocols to distinguish between them. In other

---

[†] For the monotone game, however, (and because of the nature of the game) one can get away by using pairs of vectors (x, y) such that $w(x) \leq w(y)$. Indeed, in the proof of theorem 5.1.1, we use paths which contain less than n^ϵ edges versus cuts which contain $O(n^2)$ edges.

[‡] $L = DSPACE(\log n)$.

words, the only difference between the graphs x and y is the fact that x is connected while y is not.

We conclude this work with a last open problem, perhaps the ultimate challenge and the main reason of our present endeavor. We hope that the material presented in this thesis will encourage, and help, people to work on it:

Open 6.0.7 *Show that* $C(\mathbf{Cycle}) = \omega(\log n)$.

Of course, this last problem will separate NC^1 from L solving a longstanding open problem in complexity theory.

Bibliography

[Aj83] M. Ajtai, "Σ_1^1-Formulae on finite structures", *Annals of Pure and Applied Logic* **24**, pp. 1-48 (1983).

[AB87] N.Alon, R. Boppana, "The monotone circuit complexity of boolean functions", *Combinatorica* **7**, pp. 1-22 (1987).

[An85] A.E. Andreev, "On a method for obtaining lower bounds for the complexity of individual monotone functions", *Sov. Math. Dokl.* **31**, pp. 530-534 (1985).

[An86] A.E. Andreev, "On a method for obtaining more than quadratic lower bounds for the complexity of ß-schemes",

[AKS83] M. Ajtai, J. Komlós, E. Szemerédi, "An $O(n \log n)$ sorting network", *Proceedings* 15th *STOC*, pp. 1-9 (1983).

[AUY83] A.V. Aho, J.D. Ullman, M. Yannakakis, "On notions of information transfer in VLSI circuits", *Proceedings* 15th *STOC*, pp. 133-139 (1983).

[B82] S.J. Berkowitz, "On some relationships between monotone and non-monotone circuit complexity", Technical Report, Computer Science Department, University of Toronto, (1982).

[BGS75] T.P. Baker, J. Gill, R. Solovay, "Relativizations of the $P = ?NP$ question", *SIAM Journal on Computing* **4**, pp. 431-442 (1975).

[BS88] R. Boppana, M. Sipser, "The Complexity of finite functions", Preliminary Draft (1988).

[CMcK87] S.A. Cook, P. McKenzie, "Problems complete for deterministic logarithmic space", *Journal of Algorithms* **8**, pp. 385-394 (1987).

[D84] P.E. Dunne, "Techniques for the analysis of monotone boolean networks", Ph.D. Thesis, University of Warwick (1984).

[FSS84] M. Furst, J.B. Saxe, M. Sipser, "Parity circuits and the polynomial time hierarchy", *Mathematical Systems Theory* **17**, pp. 13-27 (1984).

[H86] J. Hastad, "Improved lower bounds for small depth circuits", *Proceedings of* 18th *STOC*, pp. 6-20 (1986).

[K71] V. Krapchenko, "A method of determining lower bounds for the complexity of ß-schemes", *Math. Notes Acad. Sci. USSR*, pp. 474-479 (1971).

[K72] W.M. Kantor, "On incidence matrices of finite projective and affine spaces", *Math. Z.* **124**, pp. 315-318 (1972).

[KPPY84] M. Klawe, W.J. Paul, N. Pippenger, M. Yannakakis, "On monotone formulae with restricted depth", *Proceedings of* 16th *STOC*, pp. 480-487 (1984).

[KS87] B. Kalyanasundaram, G. Schnitger, "The probabilistic communication complexity of set intersection", *Proceedings of* 2nd *IEEE Structure in Complexity Theory*, pp. 41-49 (1987).

[KUW85] R.M. Karp, E. Upfal, A. Wigderson, "Constructing a perfect matching is in random *NC*", *Combinatorica* **6**, pp. 35-48 (1985).

[KW88] M. Karchmer, A. Wigderson, "Monotone circuits for connectivity require super-logarithmic depth", *Proceedings of* 20th *STOC*, pp. 539-550 (1988).

[MS82] K. Mehlhorn, E.M. Schmidt, "Las Vegas is better than determinism in VLSI and distributive computing", *Proceedings of* 14th *STOC*, pp. 330-337 (1982).

[MVV87] K. Mulmuley, U.V. Vazirani, V.V. Vazirani, "Matching is as easy as matrix inversion", *Combinatorica* **7**, pp. 105-130 (1987).

[Ra85a] A.A. Razborov, "Lower bounds for the monotone complexity of some boolean functions", *Sov. Math. Dokl.* **31**, pp. 354-357 (1985).

[Ra85b] A.A. Razborov, "A lower bound on the monotone network complexity of the logical permanent", *Mathematical Notes of the Academy of Sciences of the USSR* **37**, pp. 485-493 (1985).

[Ra88] A.A. Razborov, "Applications of matrix methods for the theory of lower bounds in computational complexity", *Manuscript* (1988).

[Ru80] W.L. Ruzzo, "Tree-size bounded alternation", *Journal of Computer and Syst. Sciences* **21**, pp. 218-235 (1980).

[Sh49] C.E. Shannon, "The synthesis of two-terminal switching circuits", *Bell Systems Technical Journal* **28**, pp. 59-98 (1949).

[ShS80] E. Shamir, M. Snir, "On the depth complexity of formulas", *Math. Systems Theory* **13**, pp. 301-322 (1980).

[S71] P.M. Spira, "On the time necessary to compute switching functions", *IEEE Trans. on Comp.* **20**, pp. 104-105 (1971).

[T87] E. Tardos, "The gap between monotone and non-monotone circuit complexity is exponential", *Combinatorica*, To appear.

[V84] L.G. Valiant, "Short monotone formulae for the majority function", *Journal of Algorithms* **5**, pp. 363-366 (1984).

[Y77] A. C.-C. Yao, "Probabilistic computations: Towards a unified measure of complexity", *Proceedings of* 18th *FOCS*, pp. 222-227 (1977).

[Y79] A. C.-C. Yao, "Some complexity questions related to distributive computing", *Proceedings of* 11th *STOC*, pp. 209-213 (1979).

[Y85] A. C.-C. Yao, "Separating the polynomial-time hierarchy by oracles", *Proceedings of* 26th *FOCS*, pp. 1-10 (1985).

Index

Notational Index

The MIT Press, with Peter Denning, general consulting editor, and Brian Randall, European consulting editor, publishes computer science books in the following series:

ACM Doctoral Dissertation Award and Distinguished Dissertation Series

Artificial Intelligence, Patrick Henry Winston and J. Michael Brady founding editors; J. Michael Brady, Daniel G. Bobrow, and Randall Davis, current editors

Charles Babbage Institute Reprint Series for the History of Computing, Martin Campbell-Kelly, editor

Computer Systems, Herb Schwetman, editor

Exploring with Logo, E. Paul Goldenberg, editor

Foundations of Computing, Michael Garey and Albert Meyer, editors

History of Computing, I. Bernard Cohen and William Aspray, editors

Information Systems, Michael Lesk, editor

Logic Programming, Ehud Shapiro, editor; Fernando Pereira, Koichi Furukawa, and D. H. D. Warren, associate editors

The MIT Electrical Engineering and Computer Science Series

Research Monographs in Parallel and Distributed Processing, Christopher Jesshope and David Klappholz, editors

Scientific Computation, Dennis Gannon, editor

Technical Communication, Edward Barrett, editor